Practical Hyrdogen Systems

by Phillip Hurley

copyright ©2006, 2015 Phillip Hurley
all rights reserved

illustrations and photography
copyright ©2006, 2015 Good Idea Creative Services
all rights reserved

ISBN-10: 0983784787
ISBN-13: 978-0-9837847-8-4

Wheelock Mountain Publications
is an imprint of
Good Idea Creative Services
Wheelock VT
USA

Copyright ©2006 by Phillip Hurley and Good Idea Creative Services

First print edition ©2015

Notice of Rights

All rights reserved. No part of this book may be reproduced or transmitted in any form or by any means, electronic, mechanical, photocopying, recording or otherwise without prior written permission of the publisher. To request permission to use any parts of this book, please contact Good Idea Creative Services, permission@goodideacreative.com.

Wheelock Mountain Publications is an imprint of:

    Good Idea Creative Services
    324 Minister Hill Road
    Wheelock VT 05851 USA

ISBN-13: 978-0-9837847-8-4

Library of Congress Control Number: 2015947642

Library of Congress subject headings:

    Hydrogen as fuel--Handbooks, manuals, etc.
    Water--Electrolysis--Handbooks, manuals, etc.

Disclaimer and Warning

The reader of this book assumes complete personal responsibility for the use or misuse of the information contained in this book. The information in this book may not conform to the reader's local safety standards. It is the reader's responsibility to adjust this material to conform to all applicable safety standards after conferring with knowledgeable experts in regard to the application of any of the material given in this book. The publisher and author assume no liability for the use of the material in this book as it is for informational purposes only.

Contents

Designing Hydrogen Systems

An introduction to hydrogen systems 3
Designing your hydrogen system 4
Functions within a basic hydrogen system ... 5
The system and its context 5
Effects of weather 6
Indoor systems 6
Electrical influences 6

The Design Process 9
System diagrams and lists........................... 9
The budget for your hydrogen system 10
Do it yourself or out-source 11

Working with the Materials 13
Hydrogen history and characteristics 13
Hydrogen safety 14
Oxygen 14
Oxygen and safety 15

Safety, Design and Operating Guidelines................... 17
Guidelines resources 17
Standards for hydrogen systems................ 18
Atmospheric hazard classifications 20

Process Control Safety in Hydrogen Systems 23
Safe housing of electrical and electronic process control devices 23
Pressurized electrical enclosures 23
Explosion-proof enclosures....................... 24
Intrinsically safe circuits 25
Other isolation options 27
Code requirements 27
Basic electronic, electrical, and laboratory safety 27
Electrical hazards 29
Out-sourcing circuit design....................... 29

Pressurized Hydrogen Systems ... 31
Working with pressurized hydrogen systems. 31
Protective clothing................................... 32
MAWP and MOP....................................... 32
Settings for safety release valves 33
Housing the system.................................. 33
General design safety guidelines 35
Gas concentration monitoring 35
Gas concentration values......................... 36
Hydrogen flammability 36
Hydrostatic testing 37
Fugitive emissions 38
Basic leak detection................................. 39
Leak detection devices 40
Hydrogen detectors 41
General purpose pressure design considerationsand formulas 43
Example formulas and illustration of their use 45
Pressure Regulation................................. 48
Pressure measurement conventions........... 50
Pressure unit conversion constants............ 51
Material characteristics and compatibility ... 52
Stainless steel classifications.................... 54
Brass in hydrogen systems....................... 56
Experimental vs. long term use 56
Galvanic compatibility.............................. 57
Electromotive series table 58

Building the Hydrogen System

System Overview 61
Fittings and experimental systems.............. 61
Primary subsystems................................. 64
How this system works 65

Component overview............... 67

iii

Practical Hydrogen Systems

Fittings 67
Tubing 69
Pressure tank and bubbler 69
Electrolyzer 69
Regulators 70
Storage tank 70
Gauges 71
Grades of accuracy for pressure gauges 72
Glycerine-filled vs. dry gauges 72
Gauge range 73
Coalescers, filters and driers 74
Valves 74
Adjustable and nonadjustable check valves ... 75
Accumulator 75
Equalization valve 75
Water/KOH reservoir with cover 76
Electrolyte pump system 76
Pressure switch 76
The process control unit 76
Support system 77
Tools and Materials 79
NPT connections 79
Common tools and construction materials
    for the project 80

Electrolyte Reservoir and Pump .. 83
Reservoir tank 83
Determining size of the
    electrolyte reservoir 83
Suitable materials for
    electrolyte reservoirs 84
Supporting the reservoir 84
Filter the electrolyte 85
Electrolyte reservoir cover 85
Keep the bugs out 85
Electrolyte pump 86
Pump switches 86
Pumping capacity 87
Pumps in hazardous environments 87
Check valves and the pump 88

Pump fittings should be flexible 88
Handling leaks 88
Test the pump sub-assembly 89
Setting up feed system valves 90
The electrolyte reservoir 90
Reservoir and pump parts list 92

Pressure Tank and Sight Glass 95
Pressure tank assembly 95
Assemble sight glass and connect
    to pressure tank 97
Pressure tanks and sight glass parts 99

Electrolyzer and Electrolyte 101
How the electrolyte works 101
Obtaining potassium hydroxide 101
Potassium hydroxide (KOH)
    solution strength 102
Handling KOH 103
Using KOH safely 104

Electrolyzer construction 104
Flanges 105
Alternative designs 106

Electrolyzer assembly 106
Flange ports 106
Positive electrode ports 108
Preparing the Teflon liner 109
Electrolyzer top 113
Positive electrode assembly 114
Make the silicone rubber washers 114
Make the nickel washers 114
Cut the electrode rods 115
Preparing the Teflon insulators 115
Fitting the positive electrode
    feed throughs 116
Positive mesh electrode 117
Positive electrode installation 119
Inserting the flange fittings 122
Negative electrode preparation 123
Negative electrode mesh 124

Table of Contents

Fitting the negative electrode assembly 128
Installing the negative electrode assembly .. 130

Electrode separator 134
Fitting the separator.................................. 134
Assembling the separator 135
Electrolyzer final assembly...................... 137
Attach the bottom blind flange 137
Bolt the flanges 138
Attach the top blind flange 139
Tools for building the electrolyzer............ 141
Materials for the electrolyzer 142

Bubbler Scrubber System.......... 145
Purifying the hydrogen............................. 145
Bubbler configurations............................. 145
Bubble breakers 145
Bubbler liquids 146
Additional internal filters 147
Constructing the bubbler.......................... 148
Assemble the bubbler 149
Materials for the bubbler 152

Connections and Pressure Balancing.................. 153
Hydrogen side fittings and tubing 153
Oxygen side fittings and tubing 156
Electrolyzer to pressure tank connections, bottom 158
Setting the adjustable check valve (oxygen outlet)................................. 158
Gas migration and pressure differential ... 161
The accumulator 161
How accumulators balance pressure........ 162
Calculating the hydrogen side gas volume .. 162
Calculating the oxygen side gas volume 163
Fine tuning liquid levels for the accumulator... 163
Other types of accumulators 164
Equalization valves 165
How the valve works 166
Constructing a simple equalization valve .. 167

Equalization valves with bellows............... 169
Other equalization options 170
Tools for connections & pressure balancing 170
Materials for connections & pressure balancing............................. 171

Coalescers................. 175
Coalescers vs. filters 176
Moisture in the hydrogen gas stream 176

Storage Tank and Peripherals ... 179
Pressure switch....................................... 179
Pressure switch characteristics............... 181
Installing the storage tank peripherals...... 182
Pressure switch dead bands 183
Regulator installation 184
Storage tank and peripherals parts and materials................................. 184
Storage tank connections to coalescer 186
Materials for storage tank connection....... 187

Catalytic Recombiner and Subsystem 189
Building a catalytic recombiner 189
Materials for catalytic recombiner 192
Flashback arresters 192
Source for flashback arrestors 193
Filter 193
Catalytic subsystem assembly 194
Fittings for catalytic recombiner subsystem 197

Process Controller 199
Process controller subassemblies............ 204
Process controller enclosure 206
Explosion-proof enclosures..................... 207
Pressurized enclosures........................... 207
Purged and pressurized enclosures.......... 207
Placement of electrical components 208
Specific electrical hazards 208
Relay Schematics.................................... 209

Voltage Regulator Schematics.................212
Tools and materials
  for the process controller.................214
Process controller external wire
  assembly connections....................219
Internal wire guide (process controller)....220
General wire connections.....................220
Wire assemblies.............................222
Process controller box fabrication..........223
Preparing the ports........................223
Pressure sensor............................225
Installing the pressure sensor.............226
Wire ports.................................226
Pressurize the box.........................227
Capacitive sensors.........................230
Sensor configuration.......................231
Fiber-optic sensors........................233
Other types of sensors.....................233
Sensors materials list.....................234

Hydrogen System Operation

System Setup............................239
Frame......................................239
Electrolyzer fluid level...................239

Operating Procedure....................241
Initial startup............................242
In process.................................243

Calculations for Production and Storage............................245
Measuring gas output.......................245
General voltage efficiency.................246
Calculating tank capacity..................247

Working with Commercial Hydrogen Cylinders...................249
Hydrogen purity grades.....................249
Purchase or rent the gas cylinder..........249
Storing and moving hydrogen cylinders....250
Hydrogen regulators, cylinder
  valves and outlet connections...........251

New or used regulators.....................252
Connect cylinder and regulator.............254

Nitrogen Purging........................259
Working with commercial
nitrogen cylinders..........................259
Purging apparatus..........................260
Purging techniques.........................260
Nitrogen purging sequence..................261

Hydrides for Hydrogen Storage . 265
Hydrogen storage methods...................265
Pros and cons of hydride storage...........265
Types of hydrides..........................266
Hydride bottles............................267
Charging the hydride bottle................269
Using the hydride bottle...................271
Gas cylinder and hydride
  storage resources.......................273

Commercial Fuel Cell Units........275
Mass flow controllers......................275
Purchasing fuel cell units.................275
Fuel cell unit options.....................277
Basic system inspection....................278

Suggested Reading....................287

Suppliers and Other Resources . 289

Part I
Designing Hydrogen Systems

Designing Hydrogen Systems

An introduction to hydrogen systems

The term hydrogen system can denote many things, but for our purposes it designates a system that:

- Produces hydrogen by electrolysis,
- Processes hydrogen by catalytic recombination and simple liquid scrubbing,
- Compresses and stores it in gaseous form.

Anyone can produce hydrogen by mixing a few materials, but to go beyond a simple demonstration to a practical application involves the disciplines of chemistry, electrochemistry, physics, mechanics, electronics, and electrical applications. Building a hydrogen system requires basic skills in the areas of finding material resources, understanding material compatibility, and fabricating components.

This may sound formidable, but the good news is that anyone of average intelligence, who is thoughtful and methodical, can, with a few easily learned skills, fabricate a hydrogen system at reasonable cost if they approach the process as a learning experience with the goal of developing a better understanding of these systems.

If you are a plug and play person and will hurt yourself if anything is not UL approved, you will not do well in the hydrogen lab. Patience, caution and a level head are very important, to go along with the sense of adventure that comes from the excitement of being involved in a forward looking technology.

This book is not intended to be a course in chemistry, physics or electronics. All these disciplines are touched on here as they are needed for a basic understanding of design and operation of simple hydrogen systems. We do presume some familiarity with the disciplines involved. We do not cover every aspect or possibility of design. We do work with several simple experimental ideas and develop them to finished products that will give you some idea of what is involved.

Designing your hydrogen system

Many methods can be used for hydrogen production, processing and storage, and one's choices will be determined by a variety of factors. To design and experiment with hydrogen systems, you need a good understanding of the materials and methods you are going to be working with. You will need to decide what you want your end product to be and do; and then how your hydrogen system is going to do it.

Defining your goal is usually fairly easy. Figuring out how you are going to get there is a bit more complex. It requires knowledge of components, what they do, how they do it and what type of environments they can operate in. This requires consideration of such things as pressure ratings of components; and how the materials the components are made of will react to the other materials and the conditions in the system. To avoid errors that will cause system failure or create a safety hazard, materials must be well matched and the processes of the system must be well understood.

Form follows function. First of all, you need an understanding of the primary substance you will be working with (hydrogen). Then, research the materials and components to design your subsystems; and let the system take its final shape from these.

Functions within a basic hydrogen system

A basic system involves:

- Hydrogen generation
- Hydrogen transport (within the system)
- Hydrogen processing
- Hydrogen storage
- System control
- System monitoring

All hydrogen generation and processing systems are also concerned with:

- System safety
- System efficiency – according to need or state of the art for your particular application
- Location and environment – is it to be stationary, portable or mobile, and what sort of environmental conditions will it operate in.

The system and its context

The quality (or state) of system safety and efficiency is determined by all the elements of the system and its environment converging. For instance, if your hydrogen unit will be in a submarine, different considerations come into play than if it will be in a factory setting. The environments are different, and certain space, safety and other factors are very important in one setting while they may not be relevant another. Designing a hydrogen system requires a holistic approach. You should always design a system with careful consideration of the conditions of its external environment.

Effects of weather

For instance, a simple unit placed outdoors in a stationary or mobile setting is going to be affected by the weather. Factors such as temperature changes and the rate at which the temperature changes occur can have a marked effect on materials. Some plastics, for example, do well in cold weather and others do not.

Different component materials have different rates of expansion and contraction. If they are connected, temperature changes can cause leaks if the expansion and contraction rates are too different. Some materials such as PVC are prone to ultraviolet degradation. Phenomenon like fog, dew, rain, and snow can cause electrical shorts in process control units if their housing does not protect against moisture. Most commercial units are placed outdoors with some sort of shelter to temper their exposure to ambient weather conditions.

Of course, gross weather disturbances such as high winds, lightning, hurricanes, tornadoes and flooding as well as geological phenomena such as earthquakes need to be considered in areas susceptible to any of these.

Indoor systems

If the unit will be indoors in a more controlled environment, then other factors need to be considered such as ventilation. Ventilation can be addressed through the use of gas cabinets and ventilating fans.

Electrical influences

Electrical influences on the system that should be considered are electrostatic charge, and lightning effects, whether direct or

Designing Hydrogen Systems

Grounded touchplate

indirect (such as induced charge from a nearby lightning strike). Electrostatic charge buildup can occur from blowing dust or snow, general wind conditions and a wide variety of other sources. Any metallic object (conductor) that can be electrified must be grounded so that any electrostatic charge that might build up is drawn safely to ground. If this is not done, and a charge is built up sufficiently, it can discharge to a nearby conductor – or to you – as a pathway to ground. This is similar to what happens when you get out of your car and you get an electric shock.

Hydrogen systems need to be grounded with appropriate wire or metallic braid to prevent igniting the hydrogen if there is a leak. It is convenient to have a grounded plate you can touch before you approach the system. This will avoid having a static charge your body has accumulated discharge to the metallic components, which could cause ignition if there is any leak in the system.

These are just some of the basic considerations necessary to begin to design a system.

Practical Hydrogen Systems

The Design Process

System diagrams and lists

Start the design process by making simple block diagrams of the main components of the system. Then, make a pictorial diagram indicating the exact parts you plan to use. Along with the diagram and list of parts, include the parameters of each part as it relates to your system, supplier and contact information, and the cost of each part.

Collect information resources

To develop your plan, you need to understand the components and their parts, and where to access them. You may wish to make a database of suppliers and their catalogs and reference materials. Most suppliers have online catalogs and/or paper catalogs that have plenty of information about how to use their products, including charts and tables that will help you select the products you need. You should also have trade books and other information that is pertinent to your choice of components so that you can design the system intelligently. Your database should include all cost, safety, compatibility, and performance information in order to build the system according to your experimental goal.

Once you have the information organized, you will be able to study material compatibility and identify details for your preferred system components. Consider the internal and external environment, medium characteristics, compatibility of all the materials used in the process, and the conditions of operation when studying your choices for components. Then, put your goal through the filters that will separate fact from fiction. Those filters are consideration of the cost, safety, and performance of your designed system.

The budget for your hydrogen system

You can estimate the cost of the components from your component supplier database. This will give you a general idea about whether your budget is realistic for your experimental goal. If it is not, you will need to lower the costs of components and/or reevaluate (and perhaps change) your goal.

For example, if your experimental goal is to design and build a hydrogen system to fuel a fuel cell unit for home power backup, and you found that the cost of your plan does not fit your budget, you will either have to pare down the system, find less expensive or used components, or come up with some ingenious innovation to bring the cost into line.

Every part of the system must meet safety and performance criteria. Although it is not always the case, usually safety and performance of components is directly tied to cost. Most of the time, the more expensive a component is, the more safe it is. The same is true for performance. The more costly the component, usually the better it will perform. It should be mentioned here that while more expensive components have a tendency to perform better and be safer than less expensive components, that is not always the case, and at a certain level, that may not be an important criterion. For instance, although there is a big difference between the cost of a Porsche and the cost of a Honda, both of them perform certain basic functions equally well. The Honda will not give you all the performance characteristics of the Porsche, but if you don't normally use those high performance extras, the Honda will do just fine.

Tools and safety equipment should be included in the budget for the project. The same rules apply for tools as for the components for the system. Usually the less expensive tools and equipment are not as reliable as the more expensive choice, but depending on your needs, they may be more cost effective.

Do it yourself or out-source

Another important cost consideration is to assess your own skills and what portions of the project you should out-source. For instance, if you know how to weld the particular materials you will be working with, and have access to the equipment needed, you will probably do this particular fabrication yourself. If you don't know how to weld or have the equipment, you will probably out-source your welding jobs. Call various shops and evaluate their expertise for what you need done, and get the cost for the welding job. Price and quality of work can vary widely from shop to shop. It is sometimes wise to spend more and be assured of having a professional job done as this can save you time and money later. Even if you have these fabrication skills, you may decide to out-source certain tasks.

You can learn welding and machining, and purchase equipment, if you have the time and money to do so. This can involve a significant investment for milling machines and welding equipment, and there is a substantial learning curve. However, it is always handy to have some basic machining, welding, electrical, and electronics skills; and to have the equipment to do the more basic operations yourself, even if you leave the really critical jobs to more skilled hands.

Working with the Materials

Hydrogen history and characteristics

The alchemist Paracelsus (1493-1541) mixed metals with acids and became the first person to produce hydrogen.

Hydrogen derives its name from the Greek *hydro*, meaning water, and *genes*, meaning forming, thus "water forming." Hydrogen was recognized as a distinct substance by Cavendish in 1776 and was later given the name hydrogen by Lavoisier who noticed that water was formed when hydrogen was burned.

Hydrogen is the most abundant element in the universe. Over 90% of all the atoms, and thus about three quarters of the mass of the universe, is hydrogen. This is one very simple reason why the planet is definitely going into a hydrogen economy.

Although present in the atmosphere, hydrogen is not exactly a "free floater," as it is chemically very active. It combines readily with other elements, and so is locked into compounds. On this planet most hydrogen is in water and organic compounds which make up about 70% of the earth's surface. In the atmosphere it is present at only about 1 ppm (part per million).

It is the lightest of all gases and disperses quickly if not confined. It is colorless, tasteless, odorless, and slightly soluble in water. Hydrogen can be liquefied at -423°F, and can take on a metallic state under certain conditions. At about 120.7 kilajoules per gram, it has the highest energy content of any known fuel. Its atomic number is 1, its atomic symbol is H, and its atomic weight is 1.0079.

Apart from the isotopes of hydrogen (protium, deuterium, and tritium), hydrogen occurs under normal conditions in two forms or kinds of molecules. These two forms are known as ortho- and para-hydrogen. They differ from one another by the spins of their electrons and nuclei. Hydrogen can be produced by steam reforming, electrolysis, ammonia dissociation, and partial oxidation. It can be stored for later use, as a gas, a liquid or in compounds such as hydrides. It is highly flammable and explosive, and can be easily ignited through static electric discharge; or by a catalyst such as platinum without any other source of ignition, in the presence of air or oxygen.

Hydrogen safety

Proper precautions and safety measures recommended in the hydrogen MSDS (Material Safety Data Sheet) should be followed as well as other ruling jurisdiction safety rules and guidelines when handling hydrogen. Please look up MSDS recommendations for all the materials you will work with on the internet, and study each MSDS carefully!

Oxygen

The name oxygen is derived from the Greek *oxys*, sharp, acid; and *genes*, meaning forming, thus "acid former." Priestley is generally credited with discovering it. Oxygen's atomic number is 8, its symbol is O, and its atomic weight is 15.9994. It is slightly soluble in water and becomes a liquid at -297°F. It has nine isotopes. Oxygen is about 21% of the earth's atmosphere by volume, and over 49% of the earth's crust. It is colorless, odorless, and tasteless. It reacts with all elements except inert gases, and it forms compounds called oxides. Although oxygen is not flammable, it vigorously supports the combustion of materials that are flammable. It is used in many industries for a variety of purposes. It can be produced by electrolysis, by heating potassium chlorate with a manganese dioxide, or by fractional distillation of liquid air. It is non-toxic, and as a gas poses no hazards except for its vigorous support of the combustion of flammable materials.

Oxygen and safety

Because of its support of combustion, it is important to keep oxygen separated from hydrogen, both in the production of this gas in the electrolyzer, and at any other stage of gas processing and storage. It is also important to store oxygen away from oils and greases and other hydrocarbons. Large volume storage of oxygen should be at least 20' from hydrogen tanks, or separated by a barrier at least 5' high and rated for fire resistance of at least a half hour. For connections, green color coded Teflon tape is compatible with oxygen; and LOX-8, Super LOX-8, or Oxytite are recommended pipe dopes.

Safety, Design and Operating Guidelines

There are numerous organizations who have publications that outline recommended best practices for designing and operating hydrogen systems. In the United States, these standards or guidelines are usually, at least in part, implemented by state regulatory agencies. This varies from state to state, but most rely on the standards provided by these organizations to guide them in the inspection and approval of hydrogen installations. Some states include stringent qualifications for some issues, while others will allow varying opinions about how best to proceed with safety, design and operating guidelines.

Guidelines resources

It is important to understand that these organizations issue recommendations based on professional expertise, and the recommendations, although referred to as codes and regulations, do not become law, nor are they required, unless adopted by what is called the AHJ. The AHJ is the authority having jurisdiction in your area. Most states have regulatory offices that deal with the different elements involved in building a hydrogen installation. Some states have a department of public safety with divisions for flammable gases, pressure vessel guidelines and other relevant matters. Each state is different, and you will have to refer to the state statutes to know what is required in your locale.

The particular office or person who implements these guidelines varies. In some cases, regulatory activity is a cooperative effort with many people or offices involved. State, county, or town entities may be the AHJ in your area. It is up to anyone who builds a hydrogen system to make sure they know who their AHJ is and what is required for the design and operation of their system; and to conform to all the requirements of their AHJ.

Standards for hydrogen systems

The National Fire Protection Association publications most relevant to stationary hydrogen systems are:

- NFPA 70   National Electrical Code articles 500 through 505.
- NFPA 55   Storage, use, and handling of compressed gases in portable and stationary containers, cylinders and tanks and includes some provisions for the use of hydrogen generating devices.
- NFPA 496 Design and operation requirements of purged and pressurized enclosures for equipment used in hydrogen areas.
- NFPA 853 Installation of stationary fuel cell power plants, and article 692 has some information on the installation of fuel cell systems.
- NFPA 45   Fire protection for laboratories with a discussion of allowable quantities of gas and requirements for lab ventilating systems.
- NFPA 497 Classification of flammable gases and hazardous locations for electrical installation.
- NFPA 54   The national fuel gas code.

Safety, Design and Operating Guidelines

The Compressed Gas Association publishes a variety of standards. The following are the most relevant for our purposes:

- G-5   Hydrogen
- G-5.3   Commodity Specification for Hydrogen
- G-5.4   Standard for Hydrogen Piping Systems at Consumer Locations
- G-5.5   Hydrogen Vent Systems
- H-1 and H-2   both about metal hydride storage.
- ISO/TR-15916   Basic Considerations for the Safety of Hydrogen Systems.

Another important publication is the *ASME Boiler and Pressure Vessel Code* (American Society of Mechanical Engineers). It is an expensive publication and you may wish to access it from a good reference library rather than purchase it. Most important are Section 8, Division 1 and Division 2 which contain pertinent information for the design and fabrication of pressure vessels operating at an internal pressure of 15 psig or more. ASME B 313.3 is about process piping and section 9 deals with welding and brazing.

The ASME publications cover everything you will need to know about pressure vessels, piping, and fittings for a low pressure hydrogen system. If your system is to be operated at less than 15 psig then you do not have to be concerned with these specific guidelines, although the information is well worth studying.

Important standard specifications for seamless and welded austenitic steel tubing are available from the American Society for Testing Materials (ASTM), classifications A-213 and A-249.

Some more organizations who publish relevant standards and information are ISO (International Organization for Standardization), ANSI (American National Standards Institute) and IEC (International Electrotechnical Commission). There are many documents about hydrogen systems available from the DOE (Department of Energy) and NREL (National Renewable Energy Laboratory). Other important organizations that publish material of interest are the Electrochemical Society Inc. (ECS), and the Institute of Electrical and Electronics Engineers (IEEE). All of the organizations mentioned here have websites.

If you are going to take your hydrogen show on the road, you will need to refer to the DOT (Department of Transportation) requirements, and you should review publications available from the SAE (Society Of Automotive Engineers). The SAE is actively involved in understanding and developing tight systems in a multitude of environments, thus they are a good resource for any systems designer no matter what the final application. The Department of Transportation publishes documents of interest and has a hydrogen portal website.

Atmospheric hazard classifications

Various types of hazardous atmospheres are classified to make it easier to discern the specific equipment needed to work in a given hazardous atmosphere. Most manufacturers use these classifications in their operating specifications. This classification system is clearly laid out in NFPA 497 and should be studied to get an understanding of the system. The most common classifications are Class, Division, Group, and Zone.

Hydrogen production, processing and storage would generally be Class 1, which includes flammable gases and vapors. The next classification is Division. There are two Divisions:

Safety, Design and Operating Guidelines

- Division 1 is for environments that can contain flammable gases under normal operating conditions when system maintenance occurs, or where a breakdown or failure of equipment will cause fugitive emissions.
- Division 2 is for an enclosed or confined system and where there is mechanical ventilation to remove hazardous accumulations of gas.

Flammable materials of similar hazard are classified by Group. Hydrogen belongs to Group B; so, a hydrogen production facility would be classified as Class 1, Division 1 or 2, Group B.

The code also has further designations such as Zones 0, 1, and 2:

- Zone 0 is continuously hazardous,
- Zone 1 is frequently hazardous under normal conditions,
- Zone 2 is hazardous under abnormal conditions.

In general, an experimental hydrogen system would be Zone 1. This, of course, is not always the case. Each installation or modification of an installation has to be evaluated individually.

To sum up, these classifications indicate the hazard potential for each classification and the type of equipment necessary to mitigate those hazards and operate safely. Please note that each country has its own classifications and standards which may differ from these US standards.

The standards should be reviewed not simply because you want to be in compliance, but because they will give you an understanding of why certain methods are recommended and certain are not for different atmospheres and environments. Codes and regulations are a great place to learn about the nature of the medium you are working with and the potential problems you might have under various conditions.

Process Control Safety in Hydrogen Systems

Safe housing of electrical and electronic process control devices

Electrical and electronic equipment can ignite hydrogen, even in normal operation and especially in case of malfunction. Precautions must be taken to isolate such equipment from hazardous environments such as atmospheres that may contain hydrogen/air, and hydrogen/oxygen.

Pressurized electrical enclosures

One way to do this is to use pressurized enclosures for electrical components. They maintain a slightly higher pressure inside than the atmosphere outside them, so if a leak does occur in the enclosure, it will vent outward and will not allow a flammable hydrogen and air mixture to go inside. If the pressure inside falls below a certain point, an alert occurs and the system is shut down. This can be done in different ways depending on the circumstances. Most systems have an audio alert as well as a visual indicator such as a blinking LED. In addition to this, upon loss of pressure a circuit will immediately disengage the power supply to any circuits and electrical devices within the enclosure, and the system will be shut down automatically. If you have an experimental device that is constantly attended while in the operation, a simple cutoff switch that is operator activated and approved for hazardous atmospheres will suffice.

There are several ways to pressurize enclosures. One is to apply a slight positive pressure to the enclosure and then seal it with a valve cutoff. Such a system would have a pressure read-out, or audio-visual alert, and a cutoff circuit. With this method a good seal and regular monitoring is necessary because the unit will probably have to be re-pressurized periodically to keep it at proper pressure, as very slow leaks can occur over time. Usually, this method is used when attendants or operators are present.

In another method of pressurization, the enclosure is brought up to positive pressure and is maintained by pressure sensor switch. If the pressure begins to fall, the pressure switch turns on a pump to increase the pressure. When the pressure sensor switch indicates that the pressure is at its correct level, it then cuts off the power supply to the pump. This system also uses audio-visual alerts and can use cutoffs to disengage equipment. This method is automatic and addresses the problem of the very slow leak over time; however, the alerts are necessary because an enclosure can become faulty over time either through deterioration of seals, or other problems. The frequency of the alerts and number of re-pressurization cycles can indicate to you that these issues need to be attended to.

The enclosure can also be purged of atmospheric gases and filled with an inert gas such as nitrogen at positive pressure, which will prevent combustion.

Explosion-proof enclosures

Another option for isolating electrical equipment from a hazardous atmosphere is an explosion-proof enclosure. An explosion-proof enclosure is a metal container with a flanged or threaded edge with tight tolerances, that is bolted together. An explosion-proof box does not prevent an explosion from occurring inside the box, but it quench-

Process Control Safety in Hydrogen Systems

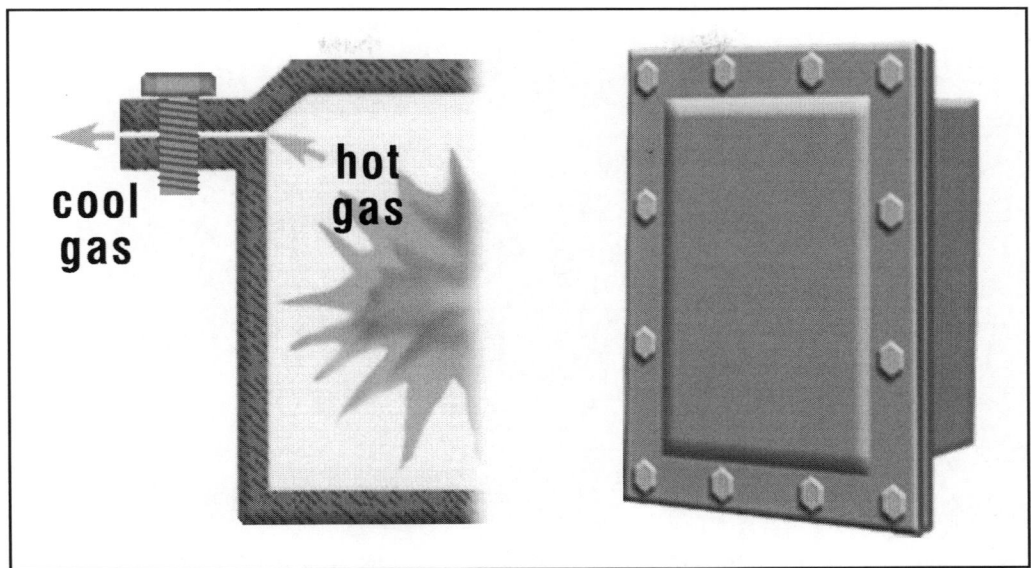

Explosion-proof box

es the flame as the flame extends outward. The flange cools the hot gases as they escape to the outside along the flange or threaded surfaces. By the time the gases exit the box, they are cooled sufficiently to not cause an explosion in the surrounding atmosphere.

Explosion-proof containers can be enhanced with pressurization or purge and pressurization methods.

Intrinsically safe circuits

Another way to address the problem of having electrical components operating in the presence of flammable gases is to use intrinsically safe circuitry. Intrinsically safe circuits limit the amount of current and voltage to a hydrogen system so that the normal operation of the circuit, or fault conditions, will not create enough energy to ignite the hydrogen in air or an oxygen rich atmosphere. Standards for intrinsically safe systems are outlined in the above mentioned NFPA articles as well as in UL 913-2002. ANSI (American National Standards

Practical Hydrogen Systems

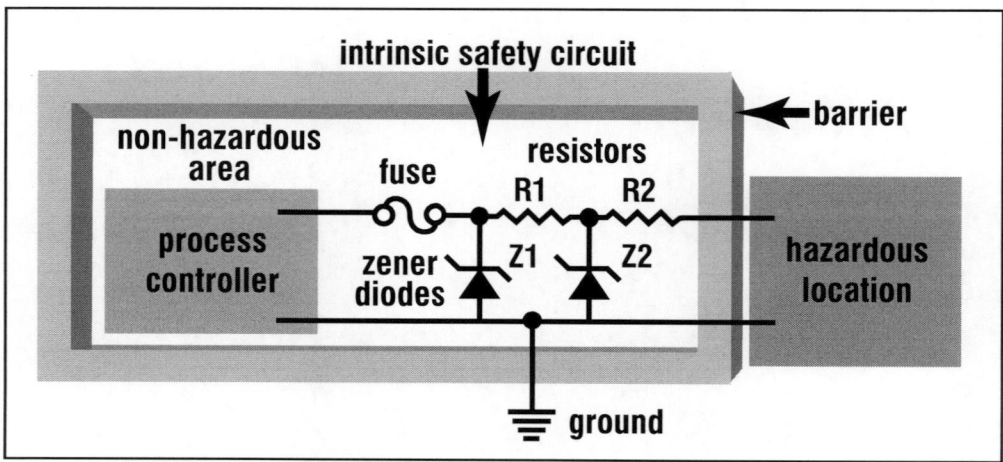

Intrinsic safety circuit

Institute) and the IEC (International Electrotechnical Commission) also publish standards for intrinsically safe circuits.

Included in the category of intrinsically safe circuits are what are considered simple circuits, sometimes referred to as simple apparatus. These are any one or more circuits that do not generate more than 1.5 volts, 100 milliamps and 25 milliwatts of power, or a passive component that does not dissipate more than 1.3 watts.

Intrinsically safe circuitry and simple circuits are used mainly for process control circuits such as pressure sensors and switches, indicator warning LEDs, and so on.

The concept of using intrinsically safe circuitry to limit current and voltage escaping into the hazardous location requires an electronic barrier that is composed of resistors (that limit current) and zener diodes that have the breakdown voltage required. When the voltage reaches a level above the breakdown voltage of the zener it is routed in another direction (to ground), thus acting as a voltage controller. Sources of generated energy for simple circuits could be solar cells that provide 1.5 volts, 199 milliamps and 25 milliwatts of power for a variety of circuits.

Other isolation options

Obviously some circuitry, electronic and electrical components are not amenable to those power limits, such as pumps and other devices which require more power. These units have to either have explosion proof or pressurized enclosures, and/or be otherwise separated from the hazardous environment. This usually means a physical barrier, and may also include distance. The ideal is to place any circuits, motors, pumps, generators, batteries, etc., in another area away from the directly hazardous area. Anything that sparks, arcs or heats up should be isolated in some way from the hazardous atmosphere.

Code requirements

Your wiring needs to meet the code requirements of your AHJ. For IS (intrinsically safe) systems, normal wiring is usually considered appropriate, but it is sometimes required that the wire be either thicker or have special insulation. Check the code for your area.

To understand code requirements and the type of equipment that meets those requirements, you may wish to have a copy of the *Cooper Crouse Hinds Code Digest* and/or *EX Digest* (which contains global standards). These guides can help you to easily synthesize and understand some aspects of electrical safety and the type of equipment required in hazardous atmospheres.

Basic electronic, electrical, and laboratory safety

In addition to reviewing all pertinent code standards and requirements, these additional steps are recommended:

- ◆ Learn the basics of DC electrical and electronic circuits and how to construct them. Even if you do not design or build your own circuits, it is important to know how they function.

- When you purchase electronics, electrical circuits, components, wire, connectors, and devices, document the voltage and current, temperature and chemical resistance ratings and your needed requirements for these items. This will give you an easy reference as you proceed with testing. The supplier and the manufacturer of the part should also be noted as different brands of components may not match in quality. Some brands may fail where other brands may not.

- Test all circuits for reliable and safe operation before final placement in a system.

- Always use appropriately rated circuit safety devices such as fuses and MCBs (miniature circuit breakers). Make sure they are placed correctly according to code.

- Design circuits that will use the least energy possible in order to perform their function.

- Make sure all components, wire, connectors and devices can carry the voltage and current necessary for them to operate and function correctly and safely.

- Make sure that all motors, pumps, make and break circuits and other devices that can spark and or arc are appropriately housed or have an appropriate enclosure or barrier in place.

- Have both an automatic power off/shutdown switch and a manual power off/shutdown switch.

- Have an appropriate fire safety plan and equipment readily available for emergency situations.

- Ground all circuits and metal components of the system.

- Make sure all lead wires are tight and secure to components and devices so that they can not be pulled away and thus spark.

Electrical hazards

Local atmospheric conditions such as thunderstorms and static charge build up are important considerations for electrical safety in your system. Static charges can build up due to windblown dust, snow, water, pollen and many other causes.

Windblown particles have been observed to build up quite a hefty charge by contact electrification on insulated bare wire fences, and in general, on any conducting surface that is isolated from ground. Static electricity is also formed by make and break contact as you exit a vehicle; or can be observed as static cling with various types of fabric. A harmless discharge to the door knob that you reach for after walking across the living room rug, can give you a little shock, but in a hydrogen hazard area, the same discharge can create an explosion. All hydrogen installations should have grounded touch plates, which people should touch before they are permitted access to the hazardous area. All metal parts should be grounded. This will safely conduct to ground any static electrical buildup. NFPA 77 discusses recommended practices where static electricity is a concern. NFPA 780 has recommendations for the installation of lightning protection devices.

Out-sourcing circuit design

If you do not wish to design your own circuits, there are electronics and electrical techs that can be found via the internet. Some of these folks can design or fabricate a circuit in an eye blink and it is sometimes better to conserve your time and energy for other tasks and rely on such resources. Many of these techs enjoy designing new circuits, or circuits for alternative energy projects and will do so for a modest fee, or sometimes even for free.

Pressurized Hydrogen Systems

Working with pressurized hydrogen systems

When you work with pressurized systems that contain flammable gases and caustic substances such as potassium hydroxide electrolyte under pressure, keep a barrier between yourself and the operating device. If fittings are not tight enough, or a component fails, the caustic KOH (potassium hydroxide) can spray out of the system. This can cause severe skin burns and blindness if it contacts your eyes. Take this very seriously and always have a Plexiglas barrier between you and the pressurized device. Always wear a face shield, protective glasses and clothing when servicing the unit or when near the unit. Protective clothing is important to avoid skin burns from spraying caustic. Also make sure that children, pets, or your cigar smoking uncle, etc., can not wander into proximity of your hydrogen system. Do not ignore these warnings as you may not get a second chance. You are responsible for safe operation of your hydrogen system, no one else.

When working with caustic, follow the advice of OSHA title 29 part 1910.151(c) which states: "Where the eyes or body of any person may be exposed to injurious corrosive chemicals, suitable facilities for quick drenching or flushing of the eyes and body shall be provided within the work area for immediate emergency use." What constitutes "suitable facilities" is outlined in ANSI Z358.1 which is the standard for emergency eyewash and shower equipment.

Protective clothing

In any area with caustic, wear full protective clothing that is rated to withstand the effects of potassium hydroxide (KOH), so that there will be a barrier between the KOH and you. Materials such as Neoprene Butyl, nitrile, and others are used for protection against caustics. Full body coverage includes coveralls, gloves, safety goggles as well as a safety face shield, and foot coverings. There are many companies that supply appropriate protective clothing and gloves, eyewash and shower stations such as McMaster-Carr, Lab Safety Supply, Industrial Safety Company, Gempler's, Consolidated Plastics.

MAWP and MOP

In general, when you work with a pressurized hydrogen system, limit the maximum allowable working pressure (MAWP) to the parameters of the component with the lowest pressure rating in the system. MAWP is the maximum allowable working pressure at which a system is designed to operate safely. So, it makes sense to consider the MAWP to be the pressure limit on the lowest rated component in the system. If a system consists of two components, one rated at 150 psi and the other rated at 1500 psi, the MAWP for the system should be 150 psi.

The significance of MAWP becomes clear when the MOP (maximum operating pressure) of the system is considered. In designing a system you decide what pressure the system will operate at. This is the MOP. The MOP needs to be 10% to 25% (or more) below the MAWP or the pressure of the lowest rated component in the system.

In determining the MAWP, we are looking at components such as fittings, piping and vessels; however, how components are fitted is also a concern in this equation. For instance, a vessel with a flange that is screwed on would have a different pressure rating than a vessel with a flange that is welded on.

Settings for safety release valves

All pressurized systems need safety release valves. The safety release pressure that is set should not be at MAWP or above MAWP but below MAWP and above MOP. Two of the types of safety release valves

Adjustable relief valves

are pop safety valves, and what are simply termed relief valves. Take a look at a supplier's catalog such as McMaster-Carr's, and study the stated difference between the different types available, then decide what type of valve you need to use for your system's pressure parameters. Please note that some valves require up to 25% over pressure before they completely open. The tolerances for safety valve opening and closing will be included in the manufacturer's specifications. Some valves have a manual pull ring for testing. Some can be set for pressure by you, or be factory set.

Housing the system

The best place to operate an experimental hydrogen system is outdoors with natural ventilation. If you are working in enclosed laboratory conditions, you may want to construct or purchase a gas cabinet with a ventilating system to enclose the gas generating and processing system. If you do this, your ventilating fans should conform to appropriate safety standards for hazardous hydrogen atmospheres. If you're working with a system in an enclosed space (with or without a gas cabinet), make sure you have the proper room ventilation. All indoor electrical components including lighting must conform to hazardous area recommendations whether you use a gas cabinet or not.

It is the nature of experimental systems to sometimes deviate from required safety standards in regard to the construction of the components. If this is the case, then it is important to use the appropriate certified barriers for any risk involved, or use certified remote operating procedures so that no personnel are injured by the operation of the device.

Low pressure systems, that is, systems that are less than 150 psi, are generally considered safer to work with compared to intermediate pressure systems with pressures from 150 psi to 3000 psi, and high pressure system with gas pressure more than 3000 psi. This is a reasonable assumption, but any lapse of safety implementation in any pressurized system can lead to disastrous results and bodily harm.

The particular system discussed in the following pages works at a modest pressure of 60 psi. However, the fact that caustic KOH under even relatively low pressure can squirt out of a system that fails and shower everything around it; and the fact that hydrogen is flammable makes the proper construction of this system within the appropriate guidelines critically important.

It is good practice to limit the amount of stored energy in a system by using the lowest pressure possible for your experimental needs.

Pressurized Hydrogen Systems

General design safety guidelines

- Make sure all components are compatible with hydrogen, oxygen and potassium hydroxide solution.
- All vessels, piping, fittings, etc., should conform to ASME and other pertinent standards.
- When you purchase components, document the rated pressure, material characteristics and other information that will be helpful for you later.
- Always work with the lowest pressure that your experimental plan will allow.
- Make operational checklists for assembling the system, for bringing it online and taking it off-line.
- Disconnect from all power sources, depressurize and drain the system before performing any maintenance.
- Post appropriate hydrogen labels to equipment and warning signs in hazardous area and do not allow entry of any but qualified, trained personnel into the area.
- Review all MSDSs (material data safety sheets) on hydrogen, oxygen, and potassium hydroxide; and post a copy of an MSDS for each hazardous material used in area at all the entrances to each hazardous area.

Gas concentration monitoring

It is good to integrate as many system failure alarm/warning systems as possible. For hydrogen leaks, for instance, a variety of different leak detectors are available. The concentration level to set a warning alarm is usually about 1% (10,000 ppm) of hydrogen by volume in air which is about 25% of the lower flammable limit, which is 4% (40,000 ppm) in air.

The LFL (lower flammable limit) is the smallest amount of hydrogen needed to ignite in air when a source of ignition is present. The opposite extreme is what is called the UFL (upper flammable limit) which is the greatest concentration of hydrogen in air that will be flammable.

After that point there is not enough oxygen to support combustion. This is 75% (750,000 ppm) hydrogen in air. Generally, the flammable range of hydrogen in air is from 4% to 75%. Any concentration of hydrogen below 4% and above 75% will tend to not ignite. There are caveats though – there are other variables that affect flammability and can change these limits.

Gas concentration values

1,000,000 ppm - 100%

100,000 ppm - 10%

10,000 ppm - 1.0%

1000 ppm - 0.1%

100 ppm - 0.01%

10 ppm - 0.001%

1 ppm - 0.000.1%

1000 ppb - 1 ppm

100 ppb - 0.1 ppm

10 ppb - 0.01 ppm

1 ppb - 0.001 ppm

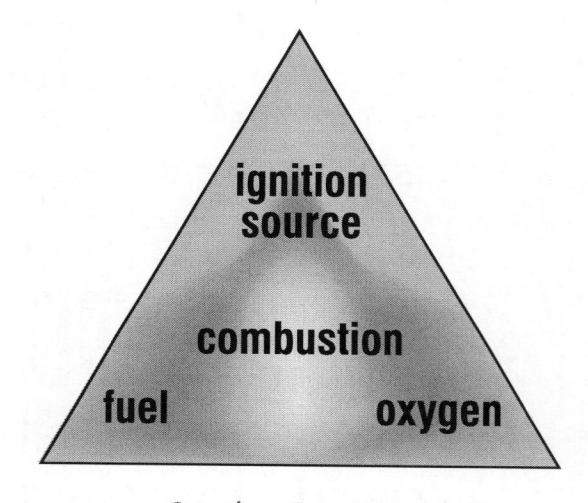

Combustion triangle

Hydrogen flammability

Hydrogen by itself is not flammable. It needs the presence of two other components to become flammable. Those two components are oxygen and a source of ignition. Without one of these components hydrogen will not burn. Oxygen is present in the atmosphere that we breath and is generated in the electrolyzer. Sources of ignition can be either mechanical such as impact, friction, fracture or rupturing and or vibration, or thermal such as open flames, hot surfaces, electrical arcs and sparks, and or catalysts such as platinum.

The good news is that hydrogen rises upward and diffuses rapidly. Unlike other types of fuel, it is quite light and buoyant and does not hang around on near ground level. However, a good wind or breeze can change all that, so it does pay to follow safety guidelines.

Hydrostatic testing

Test your pressure vessels and fitting assemblies for leaks before putting your system on line. The easiest way to do this is to use a bathtub as a hydrostatic testing reservoir. The process is very simple – pressurize each vessel or sub assembly and immerse it under water. You will see bubbles if any of the fittings are not tightened enough. Use a simple compressor unit as described later here, or something similar. The compressor should have an oil-free diaphragm if possible. Attach a pressure gauge to the compressor outlet to monitor the pressure and turn the unit off when it reaches test pressure.

You can also put an adjustable relief valve on the feed line, to release the compressed air when the component reaches test pressure. This is handier and safer than just observing the gauge and turning the compressor off when it hits target pressure. With the relief valve, you can focus on observing the leaks without having to turn the pump on and off intermittently. For this testing, the positive pressure should be at least 20% to 25% above your MOP.

Connect the assemblies to the compressor with plastic tubing and simple fittings and hose clamps. Do not immerse gauges, as some water may leak into the gauge face. These can be covered with a plastic bag and rubber bands or twist ties to keep water out; or position the components so that the gauge faces are not submerged. However, be sure that the connection between the gauge and the piping is not covered with plastic, so that you can detect leaks from this connection. You should test/observe each component or assembly for about a half hour. Gross leaks cause rapid bubbling and minor leaks will emit a very small bubble every now and then.

Monitor the compressor and do not over-pressurize the vessel during testing. Do not get the compressor connections wet as this can be a shock hazard. Keep the compressor well away from the water area by attaching an adequately long length of tubing.

Fugitive emissions

In this particular system we are concerned about two types of leaks: KOH fluid leaks and gas leaks of hydrogen and/or oxygen. Following are the main causes of fugitive emissions (leaks):

- inadequate tightening
- poor thread seal and taping
- deterioration of seals
- temperature changes which cause contraction and expansion of connections
- normal pressure cycle stress
- deterioration of components or defective component
- movement of components
- mechanical stress such as shock or vibration.

Pressurized Hydrogen Systems

Many of these problems can be avoided by hydrostatic testing before the entire system is assembled. Other important preventatives include adequate shelter for the system so that it is not exposed to extreme temperature fluctuations. Direct sunlight, for instance, will heat system components, then as the sun sets they will cool. This daily cycle can loosen the connection between components over a period of time, and can deteriorate seals. Simple shade cover with ventilation will take care of this. Seasonal temperature changes will also cause contraction and expansion of parts, and it is important to do periodic maintenance leak checks to assure a safe and optimally operating system.

Keep the pump and pump platform away from, and not connected to, any platform that supports the rest of the system. Pump operation can vibrate the components of a system and mechanically loosen them over a period of time. Likewise, be sure that anything supporting the system will not be vibrated by any other equipment.

Basic leak detection

Leak detection starts with simple observation of the pressure gauge at intervals. If during any part of the process you notice pressure decay that is not a normal part of the system's cycle, you have a leak. A visual check of the fittings and connections may show a leak if fluid KOH is involved. You can see it flowing or dripping out from the area where the fitting is connected, or see a damp spot beneath it. A hissing sound is also an indicator of a leak.

Other methods of leak detection include using soap fluid with a little water in it, ultrasonic detectors, stethoscopes, feathers and electronic gas leak detectors. The reality of leak detection is that each method has its place, but each method is limited to certain

Practical Hydrogen Systems

Ultrasonic Belfry bat detector also detects gas leaks

leak characteristics. If one type of sensor does not show a leak, another may find it.

Leak detection devices

Ultrasonic detectors are good for certain types of leaks that have enough pressure behind them to form high frequency ultrasonic waves. Sometimes I have located the general area of a leak by tracking the hissing sound, and then use the ultrasonic detector to the area to pinpoint exactly which fitting it is leaking. This is very helpful if you have a lot of fittings in a small area, and can't tell exactly which one is leaking by ear alone. There are different types of ultrasonic detectors available in the market place. I suggest using a bat detector. They're probably the lowest cost option – and they serve the purpose well. They operate at exactly the same frequency range that is needed for basic ultrasonic detection of leaks.

A turkey feather, or any good-sized downy feather, is a good visual indicator for leaks, but they can be tricky to use because they are so sensitive that

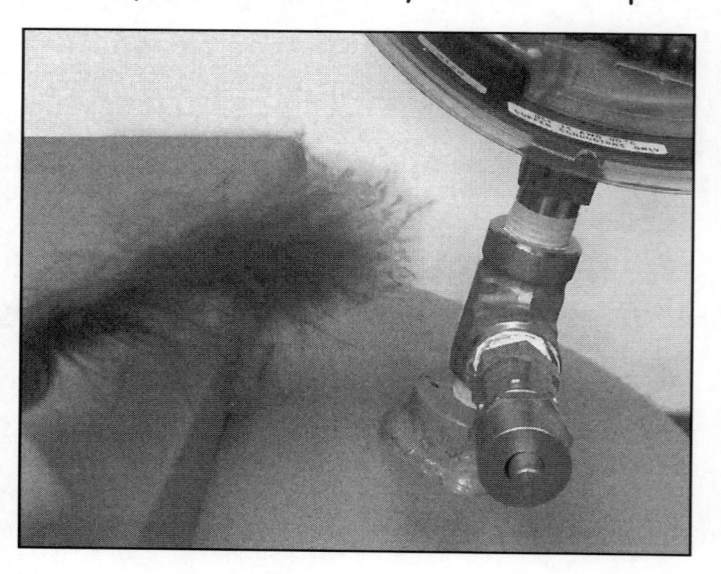

Using a turkey feather to detect leaks

any breathing or very slight movement of air from any source can also ruffle the down. It's not exactly high tech, but I have found it to be a very effective method of leak detection.

Another method is to spread a soapy dishwashing solution on the junction of fittings, and then observe it for bubbles. This works well for some leaks.

It is usually necesary to use several different detection techniques to pinpoint all your leaks.

Hydrogen detectors

You can also use a hydrogen detector as a sort of leak detector. I say sort of, because the hydrogen has to reach a certain preset concentration point before the detector indicates its presence. Although you may have a leak, that does not mean it will be at the concentration level needed for detection by the hydrogen detector. Hydrogen detectors are good for indicating concentrations of flammable gas that could be hazardous. They're good things to have installed in your system, but they should not be relied on as leak detectors per se unless you understand that they will only alert you when the a leak has reached the point of creating hazardous hydrogen concentrations.

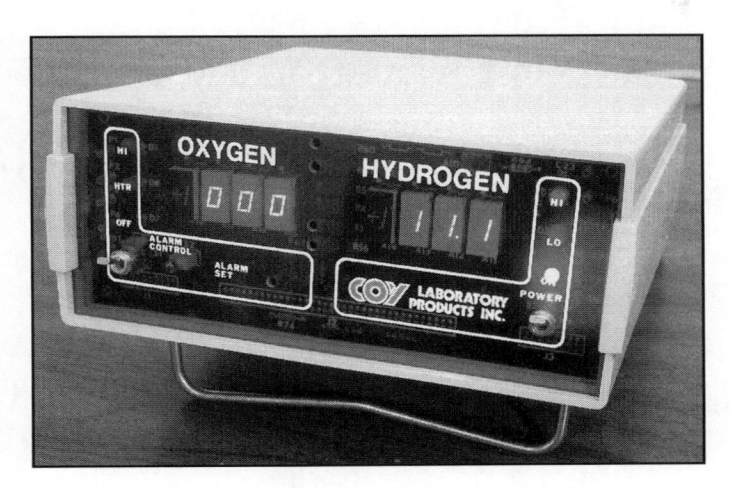

Hydrogen and oxygen analyzer

There are six basic types of hydrogen gas detectors generally available:

- Pellistor or catalytic bead sensors have a fairly long lifetime, and a wide temperature range. They require 5% to 10% of oxygen to operate.

- Semiconductor or metal oxide sensors (MOS) have a wide temperature range, but are temperature and humidity sensitive, and must be calibrated. They are not hydrogen selective and will also register other combustible gases, which can cause false alerts. This type of sensor is susceptible to short term poisoning from carbon monoxide, solvents and refrigerant gases. Sensor recovery can take several hours after such poisonings.

- Electrochemical hydrogen sensors are good down to 100 ppm, are hydrogen selective, have very low power consumption and are fairly resistant to poisoning. They operate within a narrow temperature range, require regular calibration, and usually have a shorter lifetime than other type of sensors.

- Palladium alloy sensors have a broad range of detection, and rapid response time, but they are affected by gas pressure and susceptible to poisoning.

- Thermal conductivity sensors remain stable for long periods of time and are resistant to poisoning, but they are not as sensitive as electrochemical or metal oxide sensors, and they have short lifetimes.

- Optic sensors are fabricated with fiber optic cable with coatings that react with hydrogen changing its optical properties. The light is transmitted through the optical fiber to a detector. This type of sensor is generally immune to electromagnetic influence and, depending on construction, can be safer for hazardous environments.

Pressurized Hydrogen Systems

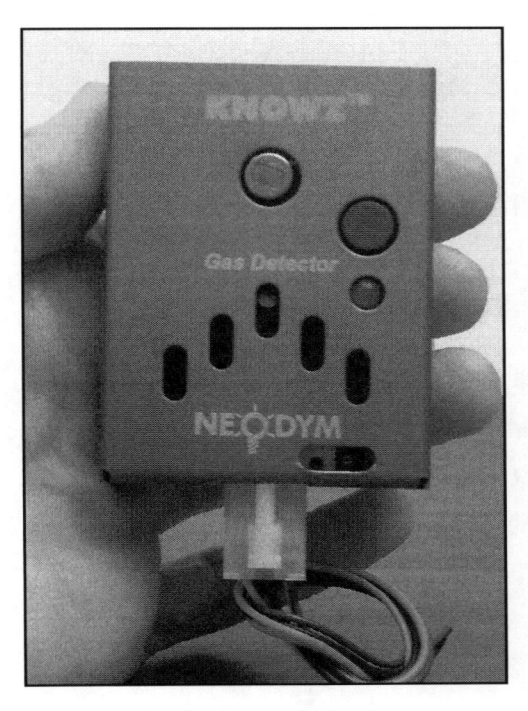

Hydrogen detector

Generally, we prefer the metal oxide detectors, despite some of their shortcomings. Much in their favor is that they are very inexpensive. For this project we used a Neodym PowerKnowz hydrogen detector for a general area alarm. This metal oxide detector is of high quality, and it's modestly priced. It can be ordered with different feature configurations. This flexibility and an easy-to-understand operator's manual make it an excellent choice for experimenters. Standard features include a bicolor LED visual indicator and interface connection. Optional features are binary (TTL), or proportional (0-5 volt) output, a beeper, relay, offset button, span trimpot, and reset button. Also, it can be powered by a 5 volt DC supply or a 7 to 40 volt DC supply, depending on which model you order. With the available options on this unit, you can set up an automatic shutoff for a system with a variety of audio or visual alarms to suit your needs.

General purpose pressure design considerations and formulas

The catalogs of most distributors of piping, tubing and fittings list the pressure ratings of their products, or at least have the information needed to figure out a product's pressure rating. However, remember that the particular way these individual components are integrated, assembled, and used is what will determine the pressure rating of the system as a whole.

Practical Hydrogen Systems

It is not possible here to give you an engineering course and present all of the information needed to appropriately design pressure vessels and systems, but the following ASME pressure vessel code sections for designing pressurized hydrogen systems provide much of the needed information. ASME Boiler And Pressure Vessel Code, Section VIII, Pressure Vessels, Division 1:

UG-23	Maximum Allowable Stress Values
UG-27	Thickness of Shells Under Internal Pressure
UG-32	Formed Heads, Pressure on Concave Side
UG-34	Unstayed Flat Heads and Covers
UG-101	Proof Tests to Establish Maximum Allowable Working Pressure
UG-125 to 136	Pressure Relief Devices
UW-9	Design of Welded Joints
App. G	Suggested Good Practice Regarding Design of Supports
App. L	Examples Illustrating the Application of Code Formulas and Rules

Other relevant publications are:

	IPT's Pipe Trades Handbook
ASME B31.1	Power Piping
ASME B31.3	Process Piping
ASME B36.19M	Stainless Steel pipe
ASME B1.20.1	Pipe Threads, General Purpose
ASME B40.100	Pressure Gauges and Gauge Attachments
ASTM A789/A789M-01	Standard specification for seamless and welded ferritic /austenitic stainless steel tubing for general service
ASTM Section 1, volume 01.01	Steel-piping-tubing-fittings

Pressurized Hydrogen Systems

Example formulas and illustration of their use

Two important formulas for your understanding are:

1. Formula for determining allowable internal pressure

$$2 \times S \frac{(T-C)}{D}$$

2. Formula for determining wall thickness

$$\left(\frac{P \times D}{2 \times S}\right) + C$$

C = allowance for threading and grooving

D = outside diameter of pipe

P = internal pressure in psi

S = maximum allowable stress value in psi

T = wall thickness

For instance, for a 304 stainless steel welded seam pipe with an outside diameter of 3½", to find the maximum internal pressure allowed for this pipe, use the pressure formula. Find what is called the "stress value" for a particular material (in this case 304 stainless steel) which is based on that material and the way it was fabricated – whether it is seamless, continuous welded, electric resistance welded, or electric fusion-arc welded. Most of the materials that you will run into will be ERW (electric resistance welded) and thus have a seam, or will be seamless with no weld. The same material of a given diameter and thickness will be able to withstand higher internal pressure if it is seamless.

For general purpose calculations I assign a stress value (S) of 18800 psi to 304 and 316 stainless steel piping and tubing that is seamless, and a stress value of 16000 to the above with an ERW (electric resistance weld, in other words, with a seam). Please note that appropriate values need to be found in manufacturer's publications or in the reference materials noted above.

For general purpose threading allowances (C) you can use .0571 for ½" to ¾" pipe, .0696 for 1" to 2" pipe and .1000 for 2½" and larger. Again, appropriate values for stress values (S) and threading allowance (C) need to be found in manufacturers publications or reference materials noted above.

Also, pipe size is an industry designation, but is not the actual measured size. For instance, a pipe size of 3" has an outside diameter of 3½". Use the actual dimensions of the pipe rather than the industry designation when doing pressure calculations.

Tubing, on the other hand, is industry designated according to the actual measured dimensions of the tube itself.

Example 1

Let's say I am looking at a supplier's catalog for 316 SS (stainless steel) tube that is 3" diameter. Seamless would be preferred due to its higher pressure rating. However, I may not need the higher pressure rating as my system is a low pressure system. Stainless steel welded tubing is less expensive than seamless and cost is a consideration. In this particular supplier's catalog only welded tubing is offered.

The supplier's catalog states that this tubing is electric resistance welded and meets ASTM A 269, has an OD of 3", and ID of 2.87", with a wall thickness of 0.065". Checking tables from the above listed publications, I note that 316 SS, ERW, ASTM A269 has a stress value of 16,000 psi.

Pressurized Hydrogen Systems

In order to find the pressure rating of this tubing, multiply the stress rating 16,000 by 2, which gives a result of 32,000. Then subtract C, which is the general allowance for threading. Since this tubing is larger than 2½", my tables tell me that the general allowance is .1000". However, I will not be threading the tube, so I do not need to subtract this from the wall thickness figure.

Next, divide T (wall thickness) which in this case is 0.065 by D (outside diameter) which is 3". Then multiply this result of 0.02166 by 32,000 which gives 693 psi. This tells me that this particular tube will withstand an internal pressure of 693 psi.

In designing systems a safety margin is added, which is usually 1½ times the operating pressure of the system. In this case I will not be exceeding 100 psi operating pressure. Calculating a safety margin of 1.5, this gives me the figure 150 psi. The tubing with an internal pressure rating of 693 psi is well above my safety margin of 150 psi, and thus is within the parameters for my system.

Example 2

To have an internal operating pressure of 100 psi as in the last example, and have the tubing be 3"OD, I need to know how thick the walls of the tubing should be for this diameter tube. First, using the safety margin of 1.5 again, gives me an internal pressure of 150 psi that the pressure ratings of the tubing must meet or exceed.

To determine the wall thickness needed, multiply P (internal pressure in psi) which is 150 psi, by D, which is 3". The result is 450. Next multiply the stress factor, 16000, by 2, which results in 32,000. Then divide 450 by 32,000 which gives a wall thickness of .014. This means that the tubes must have a minimum of .014 wall thickness.

In the supplier's catalog, all the readily available 3" tube has a wall thickness of 0.065". This is thicker than my minimum needs, but this extra thickness will leave plenty of allowance for welding factors and stress concentration.

This tubing will be used to fabricate a bubbler for an electrolyzer system. Next, I need to find some end plates that will be appropriate and can be welded onto the ends of the tube.

To reduce fabrication costs, I search for stainless steel disks. The cost of cutting would be eliminated if I could find the exact size I need. The supplier has a 3" diameter stainless steel disk, however, the smallest thickness available is ½". When fabricating ends for vessels my usual rule of thumb is to use a plate that is two times the thickness of the walls. For internal pressure, this is more than I need, but the disks will also be drilled and tapped for NPT fittings. It's good to have as much surface to surface contact as possible, within the other parameters, between the surface of the fitting and the disk. So, a ½" thick SS disk is ideal for this application. If the tubing inlets and outlets were to be welded on, rather than using screw-on fittings, it would be worth looking further for disks of a smaller thickness.

Pressure regulation

Pressure regulators are used to reduce gas pressure to accommodate a system's parameters. They are used for cylinder service or in-line service. Cylinder service requires a cylinder regulator with a particular CGA inlet to handle high pressure, and which is specific to the gas used. The outlets on the cylinder regulator will vary depending on their purpose.

In-line service requires the use of an in-line regulator. This is any regulator without a CGA cylinder connection. The in-line regulators

Pressurized Hydrogen Systems

are used at any point needed in the hydrogen generating system to reduce system pressure.

Both cylinder and in-line pressure regulators can be single-stage or two stage. Single stage regulators reduce pressure in one step. With a commercial high pressure cylinder, or a limited volume low to medium pressure tank in a system as a pressure source, the delivery pressure will increase as the cylinder or tank pressure decreases. This occurs because the decreased pressure in the cylinder or tank pushes with less force against the regulator valve, which causes the regulator to open wider.

If you are using a single stage cylinder regulator or an in-line single stage regulator that is connected to a limited volume pressure source such as a storage tank, you will need to adjust the outlet pressure periodically to maintain your target pressure.

Two stage regulators give more precise and consistent pressure. They are composed of two regulators. The first stage (regulator) is nonadjustable and reduces the pressure to a set value, such as 300 psig. The second stage (regulator) is adjustable and allows you to reduce the pressure further. Because there is less difference between the output of the first stage and the output of the second stage, these regulators maintain a steadier delivery pressure and do not require periodic adjustment.

Most in-line regulators are single stage as there is usually not much pressure difference between inlet and outlet. In-line regulators connected to limited volume storage tanks would preferably be two stage unless the maintenance of a set pressure is not a critical factor for your system.

Cylinder regulators should be two stage if periodic adjustment is not possible or convenient. The two stage regulator will give a consistent pressure delivery without periodic attention during normal operations, however, they are more expensive than single stage regulators.

The inlet and outlet pressure rating of the regulator should be compatible with your needs. If you need a specific or a certain range of delivery pressures, purchase a regulator that meets your specifications. Cylinder regulator inlets operate within common cylinder delivery pressures. For both hydrogen and nitrogen, these pressures are around 2200 psi.

Pressure measurement conventions

Psi is a common unit of pressure measure, but other units are used frequently, some of which are listed in the table on the next page.

Pressure is measured and designated by what it is relative to. For instance, to measure a pressure that is relative to absolute vacuum, this pressure measurement would be designated as *psia* (pounds per square inch absolute). Pressure relative to atmospheric pressure is designated as *psig* (pounds per square inch gauge). When pressure is relative to another pressure source, the designation would be *psid* (pounds per square inch differential). In fact *psig* is really a *psid*, however convention makes a distinction for practical use. Essentially pressure is described by comparing the pressure you wish to measure to a reference pressure.

For psig there are two distinct methods used: sealed gauge *psisg* and vented gauge *psivg*. Vented gauge is open to the atmosphere and is what would be normally used in working with equipment of the type listed in this book. Psisg gauges contain a sealed chamber with a pressure of 14.7 psi, which is atmospheric pressure at sea level.

Pressurized Hydrogen Systems

This type of gauge is sensitive to atmospheric pressure changes, and is used in altimeters and barometers.

Psi is often used when referring to pressure readings as we do here, however, it would be more correct to refer to such readings as psig or psivg.

Pressure unit conversion constants

This is not a complete list of conversion units. If you need other units, there are plenty of sites with conversion tables and software on the internet.

To go from	Multiply by	To get
bar	14.50	psi
bar	401.47	inches of H_2O
bar	100.0	kPa
bar	1019.73	cm of H_2O
cm of H_2O	0.014223	psi
cm of H_2O	0.3937	inches of H_2O
cm of H_2O	0.09806	kPa
inches of H_2O	0.036127	psi
inches of H_2O	0.2491	kPa
inches of H_2O	2.5400	cm of H_2O
kPa	0.14504	psi
kPa	4.0147	inches of H_2O
kPa	10.1973	cm of H_2O
psi	27.680	inches of H_2O
psi	6.8947	kPa
psi	70.308	cm of H_2O

Material characteristics and compatibility

This list is a small sample of material compatibility that may be of concern in constructing a hydrogen system. Many materials listed as being compatible have caveats in use. Tables are good for general outlines, but specific details of conditions of operation should be studied for each combination of materials.

C=compatible, NC=not compatible

Material	Hydrogen	Potassium Hydroxide	Oxygen
Brass	C	NC	C
Bronze	C	NC	C
303SS	C	C	C
304SS	C	C	C
316SS	C	C	C
Iron	NC	NC	NC
Aluminum	C	NC	C
Zinc	C	NC	C
Copper	C	NC	C
Monel	C		C
PCTFE	C		C
Teflon	C	C	C
Tefzel	C		C
Kynar	C		C
PVC	C	C	C
Polycarbonate	C	NC	C
Buna-N	C		C
Neoprene	C		C

Pressurized Hydrogen Systems

Material	Hydrogen	Potassium Hydroxide	Oxygen
Kel-F	C		C
Kalrez	C		C
Viton	?	?	?
Polyurethane	C		C
LDPE	C	C	
LLDPE	C	C	
HDPE	C	C	
Polypropylene	C	C	
XLPE	C	C	
Polyester		NC	
Vinylester		C	
CPVC	C	C	C
EPDM	C	C	
Epoxy	C	C	C
Silicone	C	C	C

This table indicates general suitability for low pressure systems only. You will need to consult a professional table of material compatibility and chemical resistance charts to ascertain the correctness of the above charts.

Generally, low pressure systems pose fewer problems for material compability than higher pressure systems. Epoxies, for instance, are often used to seal joints. However, the use of epoxies can be a hazard in high pressure systems because they burn, and once burning, will ignite flammable gases. Epoxies also seem to do fairly well in caustic environments, but the addition of heat and other factors can cause problems. Silicone, similarly, will do fairly well in hydrogen system environments but bears watching because the condition of use is a marked factor in its serviceability. Viton is another example of sometimes yes and sometimes no, depending on conditions of service. There are mixed readings on Viton, so it should probably be avoided. Kalrez is another example where compatibility depends on the specific Kalrez compound used.

Teflon, Buna-N, Neoprene, Kalrez and Viton need to be assessed further for use with oxygen. Teflon seems to work well in low pressure applications.

Metals such as aluminum, which perform well in low pressure applications are not recommended for high pressure application as the possibility exists for ignition in some circumstances. Aluminum, however, should not be used at all in a system that uses caustics such as potassium hydroxide.

Essentially 316SS (stainless steel) is the best material to use in a hydrogen system. Brass can be used occasionally, if it is downstream far enough from the caustic after a variety of scrubbing and filtering operations have been performed. Plated brass regulators and other components can be a cost effective alternative to stainless steel for some components. Still, if brass is used, it needs to be watched.

Stainless steel classifications

Stainless steel classifications are based on material properties, method of fabrication and use of the end product. There are various societies, institutes, and so on, that classify stainless steel according to their own concerns and uses. You will most frequently run across the classifications from the American Iron And Steel Institute (AISI) and the American Society For Testing And Materials (ASTM).

Pressurized Hydrogen Systems

The AISI classifies stainless steel into three major categories:

- austenitic which contains the 200 and 300 series,
- ferritic which contains the 400 series,
- martensitic which contains the 400 and 500 series.

Hydrogen systems which have caustic electrolytes such as potassium hydroxide need the 300 series of stainless steel, and more specifically, 304 and 316. Type 316 has a higher nickel content and contains molybdenum which gives it greater resistance to corrosion.

The L designation, such as 304L or 316L, indicates that the stainless steel has a carbon content that does not exceed .035%. This lower carbon content makes the steel easier to weld. Type 316L is the best choice for a hydrogen system. This stainless steel is most suitable for a caustic potassium hydroxide atmosphere, and not as prone to hydriding (embrittlement). Hydrogen embrittlement occurs when atoms of hydrogen penetrate certain materials and degrade their structural integrity.

304L is also a good choice. It has many of the same characteristics as 316L, but is not as resistant to caustics as 316L. It is, however less expensive than 316L.

The ASTM category for the hydrogen system detailed in this book is:

- ASTM A312 – seamless and welded austenitic stainless steel pipe,
- ASTM A269 – seamless and welded austenitic stainless steel tubing for general service.

There are other designations that should be considered for other types of service such as high pressure/high heat etc. These designations are encountered in tables used for calculating pressure allowable based on maximum allowable stress values (S-values).

Brass in hydrogen systems

As stated earlier, brass is a good material for the transport of hydrogen and oxygen within a system, but it is extremely susceptible to corrosion from potassium hydroxide. This incompatibility is a draw back, but plated brass fittings, regulators, etc. can be substituted for stainless steel, for cost savings. Also, as stated, brass can be used down-system, meaning away from emerging caustic aerosols, after the caustic has been scrubbed and filtered out of the system. It takes testing and analysis to determine the longevity of a brass device used in any particular system.

Experimental vs. long term use

For the experimental system in this book, some brass components are used, but only after filters, such as coalescers, have been installed between them and the more corrosive atmosphere directly emanating from the electrolyzer.

This system was an experimental system of short term duration to determine operating characteristics with the goal of replacing and exchanging components until the system was optimized. Thus, it was reasonable from a cost perspective to use lower cost, easily replaceable parts in some instances where caustic would not be too much of a problem.

With an experimental system, you are constantly watching the process and connecting and disconnecting components, so that you can easily view any corrosion problems that arise. However, any long term system should use at the least 304L stainless; or better, 316L stainless steel throughout; or a combination of these.

In order to set up a mature system that will behave correctly without supervision, you need to use components that are of the best quality and are totally compatible within the parameters of your system. Ultimately though, any system you design, build and operate will have to conform with the requirements of the AHJ in your area.

Galvanic compatibility

Galvanic compatibility is an important consideration for choosing materials for hydrogen systems. It is based on the relative position of metals in what is called the electrochemical series, galvanic series or electromotive series.

If galvanically incompatible metals are placed close to one another, and there is a conductive path, ions can move from one metal to the other and cause a variety of problems, one of which is corrosion.

To avoid this, choose metals that are as close to each other in the electromotive series as possible (see chart, page 58). The greater the distance between two dissimilar metals in the series, the greater the EMF (electromotive force) differential, and thus the greater their incompatibility.

It may not always be possible to have perfect compatability between different adjacent materials in the systemn, but these characteristics should be taken into account, at least, for the maintenance, monitoring and service of parts that are most susceptible to corrosion.

Corrosion causes resistance in electrical circuits and deterioration of the components involved. This results in degradation or failure of a component's performance after a period of time.

Galvanic compatibility is of most concern in electrolyzer design where a conductive KOH solution is used, and a variety of metals are used as electrodes, fasteners and containment vessels. It is also a concern for coupling dissimilar metal fittings.

For harsh alkaline electrolyzer environments, the ideal is to have no more than a 0.15v potential difference between the various metals used. In practice, however, you may have to use metals that are more galvanically incompatible than this ideal. The important thing is to be aware of this issue and its possible impact on the performance of the device in question; and in terms of maintenance required. Outside of the electrolyzer, where the environment is less harsh, an EMF differential of between 0.25 and 0.50 is ideal. Here again, this is not always attainable, but should be strived for and considered in maintenance and performance.

Electromotive series table

Material	Volts
Gold	0.00
Rhodium plate	0.10
Silver	0.15
Nickel, monel	0.30
Copper	0.35
Brass, bronze	0.35-0.45
Stainless steel	0.60
Tin	0.65
Lead	0.70
Aluminum	0.75
Iron, plain carbon and low alloy steels	0.85
Aluminum alloys	0.90-0.95
Galvanized steel	1.20
Zinc	1.25
Magnesium	1.75

As an example, the chart shows that using a low alloy steel with nickel would not be as good as using stainless steel with nickel, as the latter pair is closer to each other in the electromotive series.

Part 2
Building the Hydrogen System

System Overview

This particular system was fabricated specifically for the study, observation and experimental development of hydrogen generation, processing, pressurization and automatic control systems. It was not intended or designed for any specific application. The nature of the components chosen for this project reflect the underlying modularity and flexibility of this approach, which gave us the capability to quickly rearrange components into new configurations.

Fittings and experimental systems

Of course, the more fittings that have to be connected and tightened, and then disconnected and retightened, the more leaks there are to deal with. The downside of using lots of connectors for the sake of flexibility as you experiment is that leak management can become a major chore.

In this project we used Swagelok type connectors for the most part, but these could be replaced by quick connect fittings. This would be desirable for the ease of obtaining quick leak-free connections without the bother of tightening nuts and testing. However, quick connect fittings are more expensive.

Once a system is designed and ready to be set into final form, the number of tightened couplings can be reduced, which will reduce the cost of the system. For instance, rather than using 90° elbows, the tubing can be bent to a 90° angle, which saves the cost of the extra fitting – and more importantly, reduces the leak potential to zero in the area where the bend occurs. Similarly, any connecting point that can be welded should be. Of course, for gauge ports and other connect points where mechanical failure is a possibility (check valves, etc.), you might want to keep the ability to disengage with a Swagelok or quick connect type fitting. Any other entry

Practical Hydrogen Systems

Simple hydrogen system

System Overview

port with a fitting, such as into the electrolyzer, bubbler, pressure tank, etc., can have the fitting welded into place; or a tube welded directly to the port without the fitting to reduce cost of threading and purchasing the fitting.

Basically, start with a more flexible system, develop a design that works to your satisfaction, then review the connections to see where potential leak problems can be eliminated with pipe bending or welding, or other means. This is important if your goal is a safer, low maintenance system.

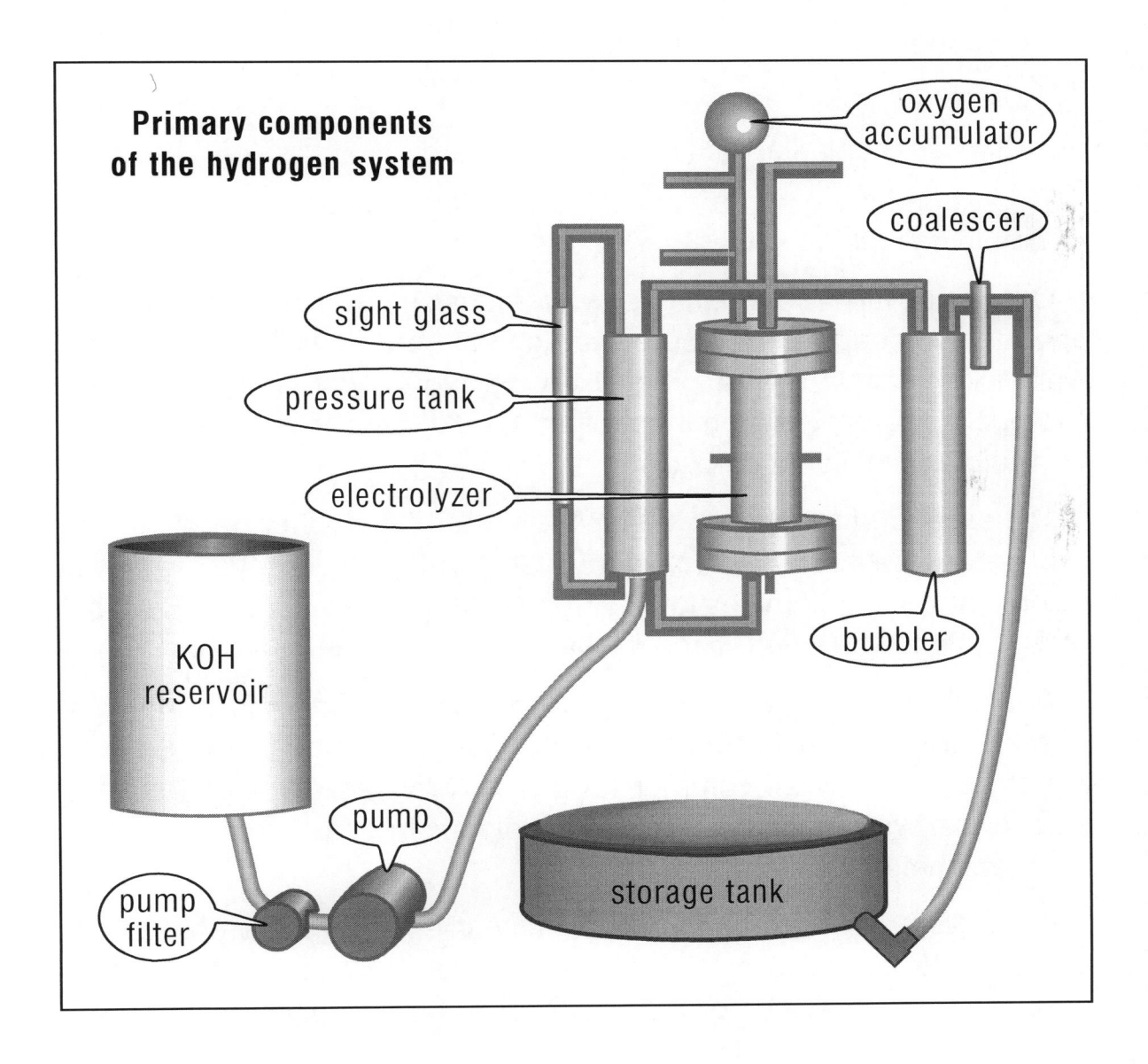

Primary components of the hydrogen system

- oxygen accumulator
- coalescer
- sight glass
- pressure tank
- electrolyzer
- bubbler
- KOH reservoir
- pump
- pump filter
- storage tank

Primary subsystems

The basic system is composed of eight subsystems:

- KOH/water fill system
- Pressure tank
- Electrolyzer
- O_2 accumulator or equalization valve
- Bubbler
- Coalescer
- Storage tank
- Process controller with sensor switches

The expanded system contains one additional subsystem – a catalytic recombining unit.

Our basic system relies on good workmanship and design, and proper operation to supply a very pure hydrogen stream. A well-designed and well built electrolyzer will produce, at the very least, industrial grade hydrogen. The only contaminants that should need to be removed are particulates and liquid aerosols.

The expanded system uses a catalytic recombiner to further purify the hydrogen gas stream from oxygen contamination. If you use a catalytic recombiner, the system has to be purged with nitrogen to move the air out of the system. The catalytic recombiner should provide a very pure gas.

Another element that could be included in the system is a chemical or mechanical dryer to further remove moisture from the hydrogen, if this is desirable. Most commercial gas is delivered dry. Gas from the electrolyzer will tend to have quite a bit of moisture in it. Coalescers and filters will help to remove moisture.

This system was designed for use with an intermittent renewable energy power supply source such as ESPMs (electrolyzer specific photovoltaic modules), wind, etc.; however, any DC power source of the appropriate current

System Overview

Electrolyzer specific photovoltaic module (ESPM) power supply

rating and voltage will work. More information about building and using ESPMs can be found in my book *Build a Solar Hydrogen Fuel Cell System*.

Our renewable energy power supply was two ESPMs that output about four volts at 20 amps (open circuit voltage, short circuit current).

How this system works

Basically the system is fed electrolyte or distilled water from a reservoir via a pump that can be manually and level sensor activated. The reservoir will supply water or electrolyte to the electrolyzer and a pressure tank with a sight glass. The sight glass on the pressure tank shows the fill level in the pressure tank and the electrolyzer, and when a predetermined level is reached, the pump is switched off, either manually or automatically with sensor switches. When the electrolyte reaches the "fill" level, the power to the electrolyzer is switched on manually or automatically, and the electrolyzer begins to disassociate hydrogen from oxygen in the electrolyte.

Inside the electrolyzer, the gases bubble upward and collect, increasing pressure. The hydrogen moves from the electrolyzer through a check valve that keeps it from returning to the electrolyzer; then it enters the bubbler. In the bubbler it is scrubbed by being bubbled through distilled water in one tank and vinegar in another tank to neutralize the KOH. From the bubbler, the hydrogen gas is fed into a coalescer which removes the fine particulates and aerosols. Then, it moves through a check valve into a tank. The tank has a pressure switch that turns the electrolyzer off when a

Practical Hydrogen Systems

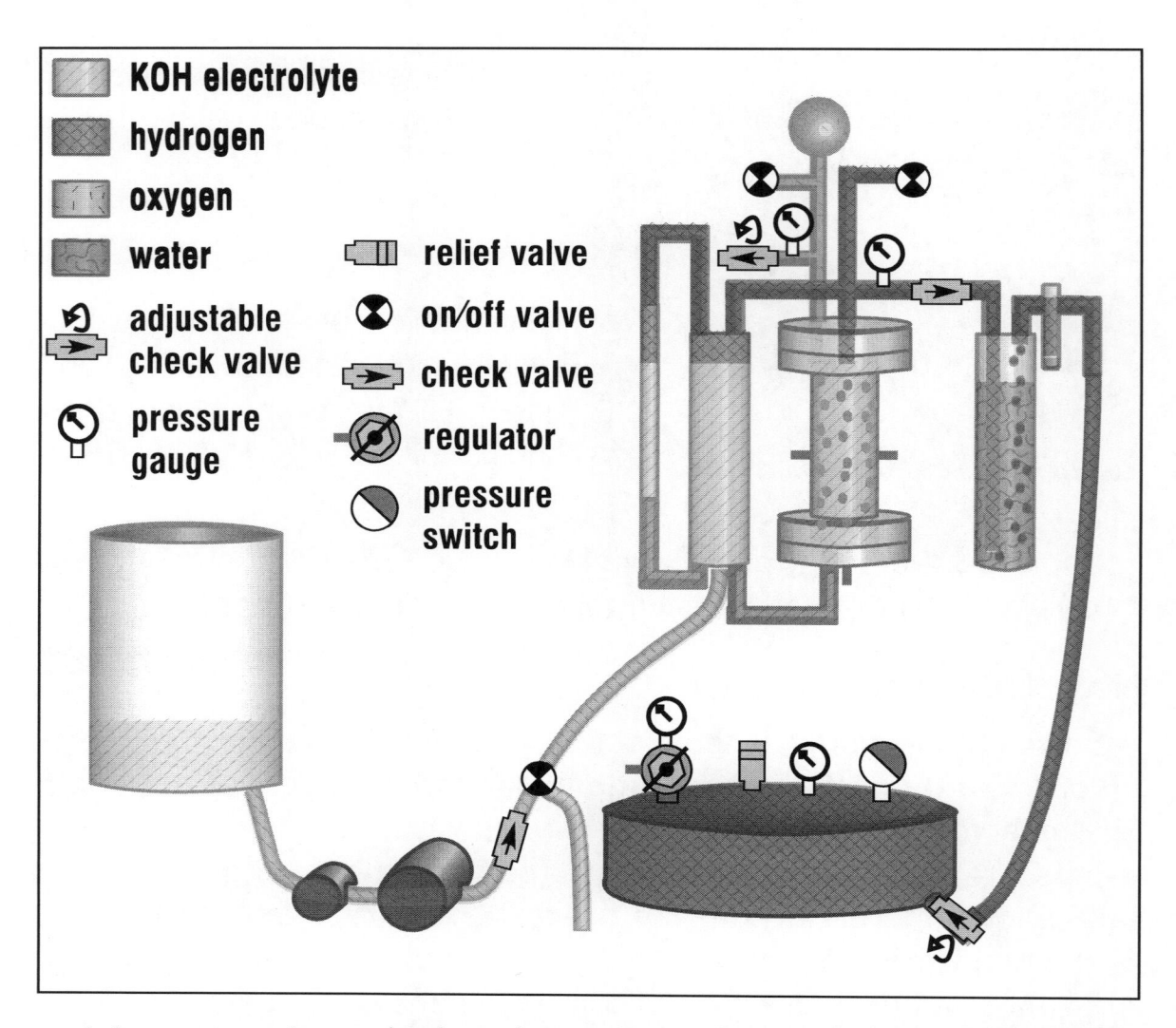

certain pressure is reached, and turns it on when the pressure falls below a certain level. It is also fitted with a relief valve so that the pressure in the tank does not exceed a set limit.

Pressure gauges are situated in line and on the storage tank for visual indications of pressure in the system.

Oxygen proceeds from the electrolyzer into an accumulator or an equalization valve to balance the pressure of the oxygen and hydrogen sides of the system.

The pressure tank pilots the pressure of the water/electrolyte pumped in to it and the electrolyzer. This is controlled (piloted) by the hydrogen pressure in the system.

System Overview

The process controller is a simple electronic/electromechanical control system. The power supply can be any DC power source, although this particular system was designed specifically to function efficiently with a photovoltaic power source.

It is basically a simple system with few bells and whistles and was designed for bare bones instruction.

Component overview

Fittings

Our intent was for this system to be modular – easily assembled and disassembled – so that a variety of sub-system configurations could be tested easily and quickly.

This was accomplished by using Yor-Lok compression tube fittings. Yor-Lok fittings are comparable with Swagelok, A-Lok, and Let-Lok type fittings.

In experimental systems, using these types of fittings is advantageous and cost effective even if the components are more expensive up front. They give flexibility in the design process which saves valuable time, and the components can be reused in other systems.

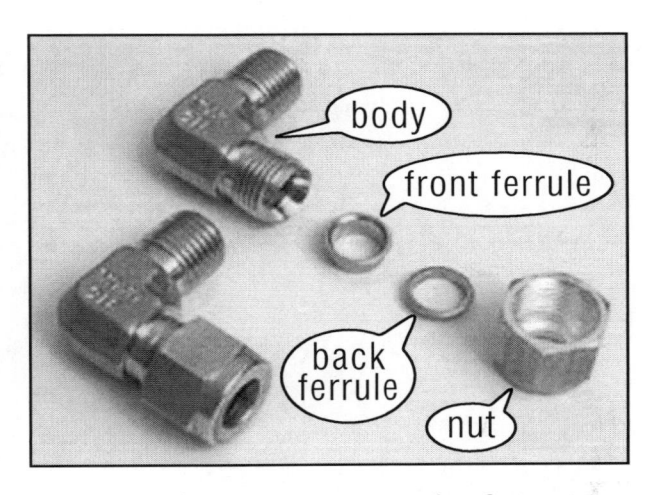

Yor-Lok compression tube fittings

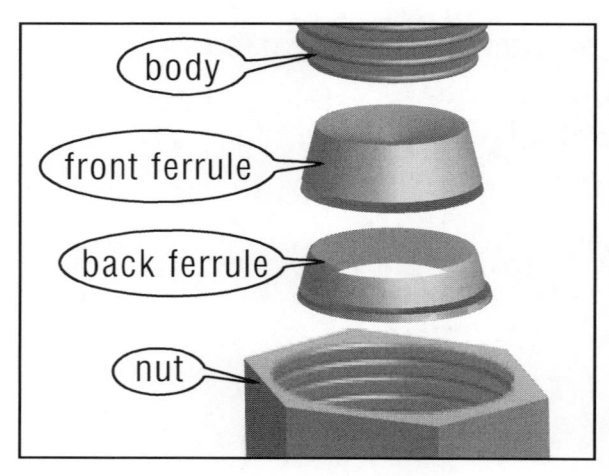

Ferrule placement in compression fittings

Below, placing and securing fittings to a tube

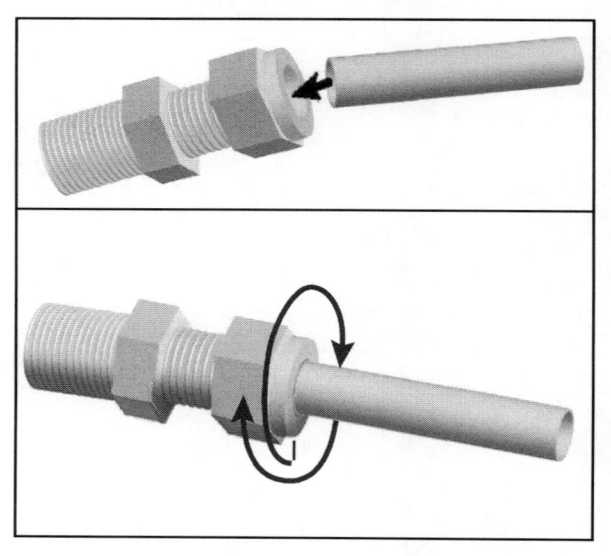

The available configurations for these fittings are basically T, cross, straight; with options for either all tube ends, or tube to NPT. The tube connecting ends consist of a nut with two sleeves (ferrules) that slip over the tubing. The sleeves are compressed and locked into the tubing by tightening the nut to create a sealed connection. For this type, do not use any thread compound or tape on the tube end connection threaded nut. You can use tape on the NPT side, however.

Some other fittings are used, as listed, and some of these are brass, but they are sufficiently downstream from the KOH to not be significant concern for a short term duration experimental project.

All of the Yor-Lok fittings used in this system are 316 stainless steel.

Pressure ratings for all fittings are listed in the distributors' catalogs or websites (McMaster-Carr, for example). Though most of these fittings will exceed low pressure rating needs, always be sure to research and double check your fittings before you buy.

For the few other general fittings that you would use in a system like this, select components with higher pressure parameters. These fittings are usually machined to closer tolerances and will provide better connections that are less prone to leaking.

System Overview

Tubing

All of the tubing in this system used for gas transit is stainless steel, except for pump hoses which are braided plastic, and a braided stainless steel hose assembly with PTFE lining connecting to the storage tank. It should be noted that stainless steel braided hose with PTFE lining is not the best to use as there is a issue with hydrogen diffusion through the PTFE. A stainless steel lining with braided steel outer covering is recommended for long term projects. Tubing OD and wall thickness can vary according to your design, as long as it meets your pressure and flow requirements.

Pressure tank and bubbler

Both pressure tank and bubbler are constructed using 3" OD stainless tubing with welded and tapped stainless steel end caps. The bubbler contains a stainless steel tube which is welded to the top cap. This tube has a stainless steel screen welded to the outlet on the bottom of the tube.

Although only one bubbler was used, this system was originally designed to have two bubble chambers. However, at least two bubblers should be used – one bubbler with vinegar to neutralize the KOH, and the other bubbler with distilled water.

Electrolyzer

The electrolyzer body consists of a threaded stainless steel pipe nipple with drilled and tapped holes for positive electrodes. The top and bottom caps are stainless steel flanges (four flanges in total) with flange seals and bolts and nuts.

The steel flanges make this electrolyzer very heavy. Another option would be to use pipe caps or disks such as used in the bubbler and pressure chamber. We used the flanges because we wanted a system that could be used for other higher pressure experiments at a later date. For a low pressure system, a much lighter electrolyzer could be constructed using, for instance, 3" or 4" diameter steel tubing with welded end caps.

Practical Hydrogen Systems

Please note that any changes to the system design have to meet pressure/safety requirements and that you are responsible to ascertain what those are for your particular project.

Remember that, although each component that goes into a system has a pressure rating, the way the components are mated needs to be calculated into the pressure rating of the system or subsystem. For instance, a screw-on flange will have a different pressure rating than a welded flange. This is because the threads in the pipe walls reduce the thickness of the wall in the threaded areas, which affects pressure parameters. Also, any break, such as a hole for electrode placement, will affect the component's pressure rating.

Regulators

The only regulator used in this system is a line regulator rated for hydrogen. It is brass and thus has to be watched for corrosion. A better regulator would be nickel plated brass or stainless steel. Because this was an experimental unit destined only for seasonal work, this was not a particular issue with us.

Storage tank

We used a portable standard steel pancake pressure tank for our tests. For seasonal and experimental use it's good because it can be inspected and flushed easily, however for a full time system you would be well advised to use a stainless steel ASME code tank. A 316 stainless tank is the best with 304 the next best.

Storage tank regulator

Your choice will depend on your budget and whether you drain and clean the tank on a regular basis.

System Overview

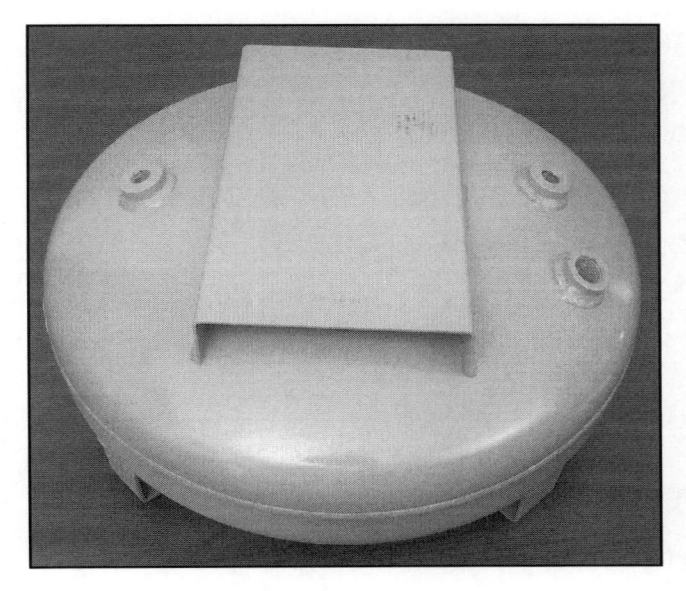

Storage tank

For testing, a small tank is desirable as it fills faster than a larger tank, so performing numerous different tests and purging is less time consuming. However, for some experiments, a larger tank might be preferable. In general, for short term use any good pressure tank will do, but keep in mind that long term storage requires materials that can weather the elements, and KOH vapors and sprays, if the tank will be exposed to these conditions.

Gauges

There are four types of pressure gauge:

- Positive gauge – a standard pressure gauge which shows positive pressure
- Negative pressure gauge – measures vacuum (pressure below atmospheric pressure)
- Compound gauge – measures both positive and negative pressure
- Differential gauge – measures the difference between two particular pressure sources.

Gauges give a visual pressure read out. They can include switches and be preset to perform a function when a certain pressure or pressure differential is reached. They are made from a variety of materials and combinations of materials, such as steel, stainless steel, brass, and plastic.

Gauges are generally designated to be suitable for corrosive atmospheres, or for general use. Gauge cases can have either back mounts or bottom mounts for connecting to tubing or piping.

Accuracies range from plus or minus 5% over full scale to plus or minus 0.1% over full scale. Of course, the more accurate a gauge is over full scale, the more expensive it is. For some applications, it is not necessary to have high accuracy.

Grades of accuracy for pressure gauges

4A	0.1% accurate full scale (the most accurate)
3A	0.25% accurate full scale
2A	0.5% accurate full scale
1A	1% accurate full scale
A	1% accurate mid scale and plus or minus 2% accuracy over the first and last quarters of the scale.
B	plus or minus 2% accurate mid scale and 3% in the first and last quarter of the scale
D	generally plus or minus 5% accuracy over full scale.

Understanding gauge accuracy designations is important, as errors can be introduced into your experimental work if you do not take into account the accuracy of the reading. Accuracy is important, for instance, for setting adjustable pressure equalization valves, so you will want to use as accurate a gauge as possible.

Glycerine-filled vs. dry gauges

Gauges can be either dry or glycerine-filled. For the most part, dry gauges would be used in this type of system. Glycerine-filled gauges are used in areas where there is a lot of vibration, as the glycerine reduces needle jump caused by vibrating components in the area. We use a small lab compressor for a pressure source for hydrostatic testing and adjusting valve pressures, and since the compressor vibrates quite a bit, we use

System Overview

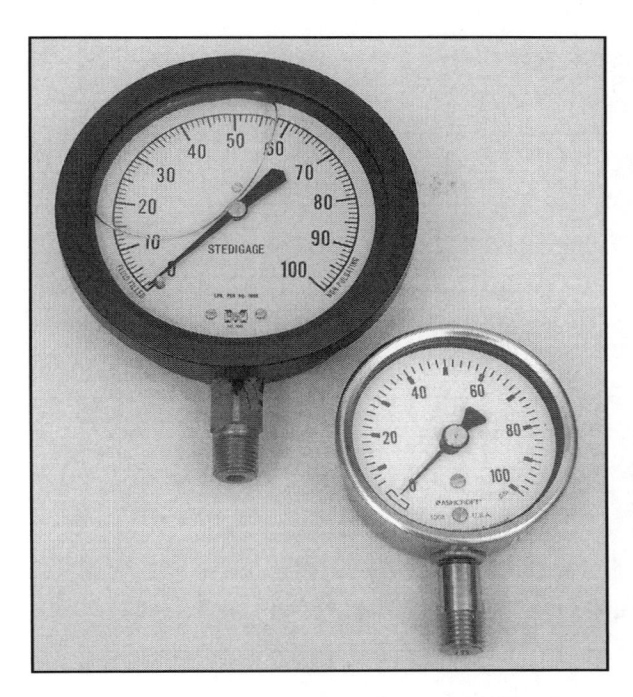

Pressure gauges, glycerin-filled (left) and dry (right)

glycerine gauges for these purposes. Glycerine gauges are also good for pump systems, which tend to vibrate, if pump pressure readings are desired.

Gauge range

The face of the gauge has figure intervals and graduation marks that indicate the pressure or vacuum being read. Figure intervals are numeric indicators such as 2, 10, 20 and so on. The graduation marks are the lines in between the figure intervals that have no accompanying number signature. Their range can vary, that is, graduation marks can indicate 2 psi or 5 psi or 10 psi intervals. In choosing gauges, be sure the figure and graduation marks reflect the accuracy of the readings needed to accomplish your particular task.

Each gauge also has a pressure range. It is recommended to use a pressure gauge that has twice the pressure range of your intended normal operating pressure, or at least your MOP (maximum operating pressure) should not exceed 75% of the full scale range of the gauge.

Gauges should be stainless steel throughout. Beware that some gauges are stainless steel on the outside, but not on the inside. Do not use brass gauges. It is most critical that what is called the wetted part of the gauge be stainless steel, as this is the part that comes into contact with the medium. In the case of this type of system, there are KOH vapors involved that will deteriorate brass gauges.

If the system will be evacuated before a nitrogen purge, install compound gauges that will show both vacuum and pressure.

Coalescers, filters and driers

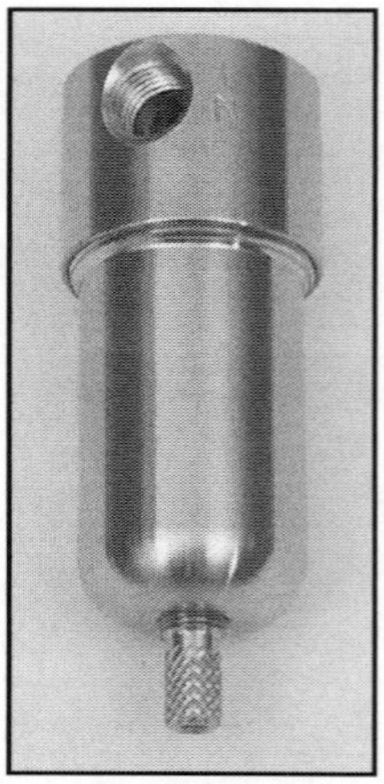

This system has a stainless steel coalescer, and in another configuration, a brass (polyethylene or urethane) filter. For seasonal use, the brass/poly is fine, but for long-term use, another stainless steel filter would be the choice.

The coalescer removes particulates and aerosols. I recommend using a larger coalescer, and also adding another water filter and a gas dryer to try to remove as much moisture as possible. We found that the system in field tests carried a lot more water than expected. The storage tank collected much of this moisture, and although this not a big problem, it is not desirable.

Coalescer

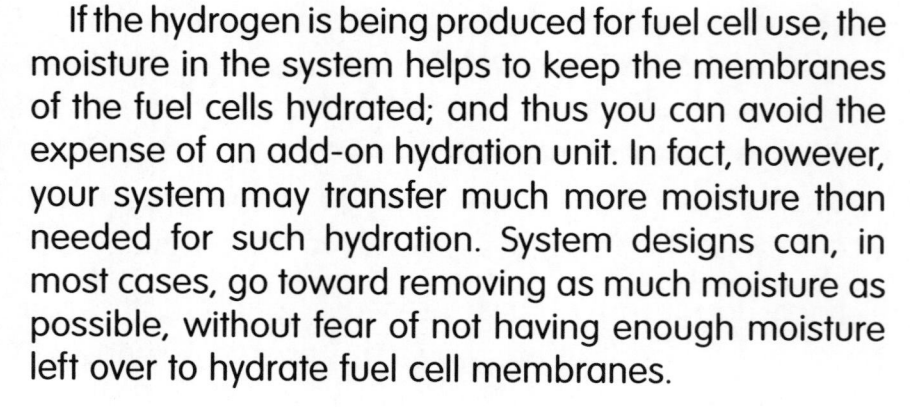

If the hydrogen is being produced for fuel cell use, the moisture in the system helps to keep the membranes of the fuel cells hydrated; and thus you can avoid the expense of an add-on hydration unit. In fact, however, your system may transfer much more moisture than needed for such hydration. System designs can, in most cases, go toward removing as much moisture as possible, without fear of not having enough moisture left over to hydrate fuel cell membranes.

Filter

Valves

The two main valves used in this system, located on the hydrogen and oxygen outlets, have Yor-Lok tube connections. More valves could be added to the system, depending on your system parameters.

A PVC valve can be located between the pump and the pressure tank for draining the electrolyte.

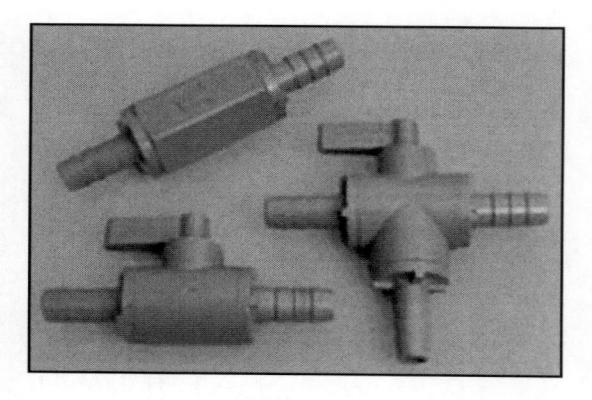

PVC valves

System Overview

Adjustable and nonadjustable check valves

Our design uses one check valve between the bubbler and the electrolyzer to prevent back-flow of hydrogen into the electrolyzer from the bubbler.

There is also an adjustable check valve functioning as a gas release valve at the oxygen end of the system, and another located at the entrance to the hydrogen tank functioning as a release valve into the tank.

Above, adjustable check valves; below, non-adjustable check valve

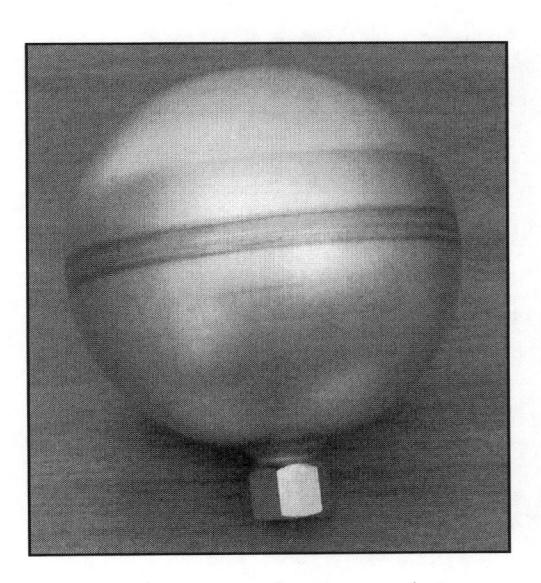

Stainless steel accumulator

Accumulator

One option for balancing the oxygen and hydrogen accumulation and pressure within the system is to use a storage equalizer called an accumulator.

Equalization valve

Another option for balancing the oxygen and hydrogen accumulation and pressure within the system is an equalization valve. The particular valve fabricated for this system consists of a piston with attendant machined fittings.

Water/KOH reservoir with cover

Any holding tank made of a KOH compatible material will do for the water/KOH electrolyte, with a feed to the pump via braided plastic hose.

Electrolyte pump system

We used a simple on-demand 60 psi Flojet pump with pre-filter to screen particulates. Pump system activation can proceed from level indicator and or manual feed and cutoff.

Fill pump and filter

Pressure switch

The tank is fitted with a bourdon tube mechanical pressure switch, which sends a signal to the process controller to turn the electrolyzer on or off.

Pressure switch for storage tank

The process control unit

The process controller routes various signals for process level and pressure controls. The unit itself is contained within a pressurized box. The box has its own pressure indicator.

System Overview

Process contoller completed and operating

Support system

The system needs some type of support. This can be designed and fabricated from a variety of materials in whatever configurations suit the requirements of your system.

Tools and Materials

At the end of each of the following sub-assembly sections there are lists of tools and or materials needed for that particular subsystem. There are some general tools and materials not listed which are used to fabricate more than one sub-assembly. In order to avoid redundancy I will list some of these items here.

NPT connections

For all NPT connections you will need thread sealant and or thread tape. We use both; however, it is not necessary to do so. Use the yellow military grade tape and a high quality thread sealant, as listed in the table that starts on page 96.

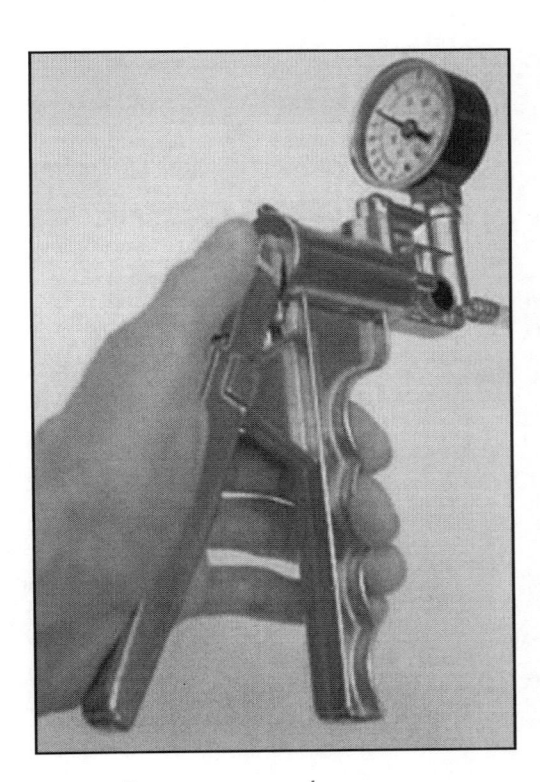

Pressure and vacuum hand pump

Wrenches

An assortment of different size wrenches are needed, and in many cases, two wrenches of the same size, to get tight fits throughout the system. It is, of course, best practice to have all the correct sized wrenches so that you don't ruin the nuts and bolts.

Miscellaneous tools

Other items such as tin snips, regular pliers, needle nose pliers, razor knife, scissors, rulers, T-squares, Sharpie markers, wire connection crimpers, and wire cutters will be needed. Hand punches and reamers are also used.

Mill or drill press

To mill the holes and tap the various components requires a mill or drill press. The mill ends and taps needed for fabrication of components are listed. If you have your local machine shop do it, obviously you will not need these items.

Tube cutting

To cut tubing, you will need a cut-off saw with a steel cutting blade as listed in the tool list, and V-blocks with clamps to hold the tubing tight while cutting. A hacksaw can be used to cut the tubing, or the pieces can be cut at a machine shop.

A bench grinder is handy to smooth out burs on the edges of cut tubing, but a hand file can also be used instead.

Common tools and construction materials for the project

Description	Supplier	Part #	Quantity
Military grade thread sealant tape, premium, colored. One roll each of ¼", ½", ¾", 1"	McMaster-Carr	4494K11, 4494K12, 4494K13, 4494K14	One roll of each
PTFE oxygen-safe joint sealant 3.5 oz tube, USDA approved, -400° to +500° F	McMaster-Carr	4538K1	1
Epoxy in plastic dispenser bottles, high peel strength, gray, 8 oz	McMaster-Carr	7508A43	1
Multi-meter	electronics supplier		1 or 2
Center square	MicroMark	82285	1

Tools and Materials

Description	Supplier	Part #	Quantity
Hand-held vacuum/pressure pump zinc alloy head, 25" hg maximum vacuum, 15 psi	McMaster-Carr	9963K21	1
High-performance cutoff wheel 60 grit, 10" diameter, .040" thick, ⅝" arbor. A smaller size wheel can be used	McMaster-Carr	45615A148	1
Milling vise	Micro-Mark	82747	1
V-block match set 2/clamp	Micro-Mark	14255	1
Milling machine or drill press	hardware store		
Cutoff saw or hacksaw	hardware store		
Plastic tubing, variety of sizes	hardware store		
Piston compressor/vacuum pump 20" hg max vacuum, 90 psi max pressure	Herbach & Rademan		

Parts numbers and suppliers have been provided for your convenience; however, suppliers may go out of business and parts numbers may change. All parts listed are available from multiple suppliers.

Electrolyte Reservoir and Pump

Reservoir tank

The reservoir tank holds distilled water or KOH solution electrolyte in reserve to replenish the electrolyte in the pressure tank and electrolyzer. The electrolyte fluid is depleted when the electrolyzer operates and water is disassociated into hydrogen and oxygen. There are also losses through aerosol migration.

Since the system is pump fed, the reservoir can be above, at level, and/or below level, relative to the rest of the hydrogen system. Generally, for a system like ours, the reservoir tank can be any three to five gallon tank that has a barbed outlet. The barbed outlet is connected to a hose, which connects to the pump inlet. The barbed outlet can be either on the side near the bottom of the tank, or directly on the bottom of the tank.

Electrolyte reservoir

Determining size of the electrolyte reservoir

The tank size can be under 3 gallons or more than 5 gallons, if some other size suits your needs better. The reservoir tank is filled on an occasional basis, as needed. Operating the system will give you a better idea of what tank volume is optimal, based on how much fluid is used up and how often you want to fill the reservoir. If a system is set up in an unattended remote location where maintenance and observation visits are infrequent, a much larger reservoir may needed. Or, in a very sunny location when the electrolyzer is powered by photovoltaics, water will be used up much more quickly than in a region that is only partly sunny for most of the year. If the system is being run full time on rectified DC from the mains, a larger reservoir is also needed. These kinds of considerations will determine the optimal size of the electrolyte reservoir for a given system.

Suitable materials for electrolyte reservoirs

The material that the reservoir tank is made of must be KOH resistant. Some appropriate materials include:

- LDPE (low density polyethylene)
- LLDPE (linear low density polyethylene)
- HDPE (high density polyethylene)
- XLPE (cross linked polyethylene)
- PP (polypropylene)
- PVC (polyvinyl chloride)
- CPVC (chlorinated poly vinyl chloride)
- Ryton (polyphenylene sulfide),
- EPDM (ethylene propylene diene monomer).

Stainless steel is the only metal that can be used for a reservoir tank for KOH solution. Never use aluminum tanks or utensils when working with KOH solution.

The basic five gallon buckets available at most hardware stores make good reservoirs. Just drill a hole in the bottom or side, sized to accept the NPT side of a NPT/barb fitting, then glue it in with epoxy.

Supporting the reservoir

If the hole is in the bottom of the bucket rather than the side, the bucket will have to be supported so that the hose can connect to the fitting. For bottom drain reservoirs of this size, at most discount stores you can find a plastic cube crate (like milk crates). The hose can fit through any of the spaces in the crate wall, and the crate provides solid support for the reservoir. If the outlet is on the side, and high enough for the hose and fitting to clear the ground/floor, you don't need to elevate the reservoir.

Electrolyte Reservoir and Pump

Filter the electrolyte

Wherever the outlet is, a strainer/filter of at least 40 mesh screen should be placed over the outlet on the inside of the reservoir and epoxied in place. If a pre-pump filter is placed just before the pump in the system, this is not necessary. However, the screen is less expensive than buying a pre-filter, and it is easy to make.

Tanks can be purchased from such suppliers as McMaster-Carr, Aquatic Ecosystems and US Plastics. These may already have outlets, or not.

Electrolyte reservoir cover

The top of the reservoir should have a screen cover to keep out bugs and other debris, and should be covered by a hard cover with holes in it. The holes equalize pressure as liquid is pumped out of the reservoir. Five gallon buckets come with lids which can be drilled for air holes. Then, place the lid over a piece of screen material that has been laid over the bucket top.

Keep the bugs out

Although this may seem like an insignificant item, be aware that if an insect gets into the system, it can jam up things enough to require draining and flushing, and possibly taking things apart. A screened reservoir outlet can prevent a lot of pain and aggravation. As a matter of course, tubing ends and openings in the system should be taped over when the system is not running. Insects (earwigs especially) like to crawl up hoses and into piping and fittings.

We have used many types of containers for reservoirs. They can be easily and inexpensively fabricated. For this particular project we used a 10" PVC pipe cut off from a construction site and glued it to a PVC bottom plate, into which we drilled a hole and added a barb fitting.

Electrolyte pump

The pump in this system is a FloJet model 2100-732, 12 volt, 10 amp, 60 psi, intermittent duty, demand pump. This is a self-priming diaphragm pump, and thus can be run dry without harm.

The pump's purpose in this system is to move either distilled water or KOH from the reservoir tank to the pressure tank and electrolyzer at a pressure which is slightly above the operating pressure of the system.

This FloJet can be operated from a 12 volt battery which is charged by a photovoltaic panel, or some other source of DC (direct current) electricity. Since the pump is rated at 10 amps, the power supply must provide 12 volts with at least 10 amps. The operation of the pump is intermittent and only for very short periods of time.

Electrolyte pump

There is a strainer/filter available for the FloJet that is placed just before the pump inlet, or a homemade strainer/filter can used in its place. The strainer should be at least 40 mesh stainless steel screen. If a strainer/filter has been installed on the outlet of the reservoir tank as mentioned previously, this is not necessary.

Pump switches

The pump is controlled by a manual switch and a sensor switch. The manual switch can be used to initially fill the pressure tank and reservoir. After the initial fill, the automatic sensor switch takes over to maintain a predetermined fill level that can be seen in the pressure tank sight glass. The fill sensors are positioned to register the minimum and maximum fill levels on the sight glass.

Electrolyte Reservoir and Pump

The electrical connections are simple and straightforward, as shown in the diagram/schematic on page 249. The manual switch is a SPDT 35 amp toggle switch. One side of the switch is spring loaded to center and the other side is maintained until manually pushed off. This makes it possible to use the spring loaded side for a short, quick jog, or the other side to have the pump run continuously. The jog side is a refinement that allows a very quick spurt when just a tiny amount of additional electrolyte is needed to bring the level up to fill level. It is also useful for getting used to filling a system, so that you can see in increments how fast the pump delivers electrolyte to the system.

Pumping capacity

Our pump is rated at this capacity:

GALLONS PER MINUTE	SYSTEM PRESSURE
1.5	0
1.4	10
1.2	20
1.1	30

GALLONS PER MINUTE	SYSTEM PRESSURE
0.9	40
0.3	50
0	60

Pumps in hazardous environments

This pump worked out quite well in our system. However, it is not rated for use in an explosive environment, so I cannot recommend using it. We distanced the pump from the system, and placed it at a lower position than other system components. Hydrogen rises swiftly and dissipates when not confined, so a lower and distanced placement of this equipment satisfied our safety requirements. However, it may not meet the standards necessary according to your AHJ.

This particular FloJet could be enclosed in a positive pressure vessel, as was done with the process controller for this project. It could easily be slipped into standard PVC pipe fitted with screw-on pipe caps and a pressurization port. Also, an air driven pump and/or pulley driven pump could be used instead, or purchase a pump that is rated for hazardous environments.

We had this pump on hand from a previous solar project. It suited the system's requirements for pressure delivery, and it could be powered by a PV/battery system, which was an important criterion for us.

Depending on the pump you use, make sure that it is shielded from sunlight, dust, dirt and water, including wash down spray, rain, snow, heavy condensation, and dew.

Check valves and the pump

If a check valve is used in the feed system, it should not exceed 2 psi cracking pressure, as it can interfere with the priming ability of the pump.

Pump fittings should be flexible

It is also important to use flexible hose, and plastic inlet and outlet fittings. This is because pumps vibrate, and rigid connections do not fare well in this type of environment. Rigid metal fittings can be used at the pump, but if they are tightened too much, they will tend to crack the plastic inlet of the pump, during either installation, or during operation from vibration. In general, rigid pipe and fittings should be avoided for this purpose.

Handling leaks

Most likely, if leaking is going to occur in the liquid part of the system, it will occur at the pump. Thread sealant or thread tape can be used, but do so sparingly. The reason is twofold: particles of thread tape or thread sealant can inhibit valve action causing priming problems; and tape or sealant makes it easier to crack the housing when installing fittings. If you use thread tape, do one wind only. If you encounter a leak during opera-

Electrolyte Reservoir and Pump

tion, tighten the connection little by little and observe. Don't try to tighten in one big step as you will be more likely to crack the housing.

Test the pump sub-assembly

Before connecting the pump sub-assembly into the system, set up the pump with hoses connected on the inlet and outlet and turn the pump on, feeding it water. Observe the action of the pump to get a feel for it and notice if there are any leaks. Such a test is limited because there will not be back pressure from the whole system (unless you provide for that), but this test will indicate any gross leaks.

When our system was first set up, there was a leak at the pump outlet. It was not a major leak, but it was significant enough to have to be fixed. We decided to wait a few days; however, after a few days the leak had stopped. We have observed this same phenomenon in several of our earlier systems. What may happen is that the pressure within the system forces the tape and or thread sealant to form a barrier and thus stops the leaking. There was also some crystallized KOH around the surface of the leak area, so this may have also figured in the leak healing phenomenon. Whatever!!!

Please note that if a system exposed to freezing temperatures, it must have an adequate amount of KOH in the line to act as an antifreeze; or drain the line and pump so that it is liquid free. If the system will not be used in the winter and it is in an outdoor location, it should be drained and flushed. We drain and flush our systems on a seasonal basis. This allows us to clear out the system and make any needed repairs.

An experimental system may need to be drained more than just seasonally. It's always wise to design systems to drain easily and efficiently.

Setting up feed system valves

The valves in the feed system can be set up in a number of different ways.If a check valve is used to keep the KOH from reentering the tank, put a check valve between the pump and the pressure tank, right after the pump. Beside the check valve on the line to the pressure tank, place a 3-way valve. This valve will allow the fluid from the pump to proceed either to the pressure tank and electrolyzer; or be drained from the pressure tank and electrolyzer to an emptying container. After the electrolyzer and pressure tank are drained, the fluid can be pumped out of the reservoir, if desired. This will completely drain the system of KOH solution/distilled water.

The 3-way valve can also be used to lower the KOH level in the sight glass, if necessary. If the liquid level is above the fill point in the sight glass. open the valve to release some of the KOH until it reaches proper level.

For a gravity drain of the pressure tank and electrolyzer as mentioned above, the pump and 3-way valve must be located below the electrolyzer and pressure tank, so that the fluids will drain downward. The reservoir can positioned at any level, since the pump is used to drain it.

The electrolyte reservoir

Fabrication of the reservoir/pump system consists of choosing a reservoir. If the reservoir does not have an outlet, you will have to fabricate one by drilling an appropriate sized hole to snugly fit the NPT end of a NPT

Electrolyte Reservoir and Pump

barb hose connector. Since the walls of most containers will be thin, tapping is not necessary, but will give a tighter fit. It is usually sufficient to drill a hole a little smaller than the fitting thread, then inset and screw the NPT end into the container. Then, apply epoxy to the edges to seal.

The NPT/barb hose fitting should have a barb end the same size as the barb end for the inlet strainer, if used; or the same size as the barb inlet on the pump so that one piece of hose can be used without adding an adapter.

Our FloJet pump came with a barb inlet and outlet to fit ½"ID tubing. The inlet strainer also had a ½" inlet and ½" outlet. Thus, we used a ¼"NPT x ½" barb connected to the bottom of the reservoir and connected a ½"ID Tygon braided hose to the barb connector with a hose clamp.

Hose clamps should be used to secure fluid distribution tubing, but do not over tighten, especially if using the typical worm-drive hose clamp. The barb on the barb connectors is what seals the tubing, and the hose clamp just keeps the hose from being pulled off the connector. The worm drive hose clamps have slots along the bands, which, if tightened too much, will cut right into the hose and compromise the hose's strength. They are fine to use as long as they are not over tightened. The hose clamps should be 316 stainless steel.

Reservoir and pump parts list

Description	Supplier	Part #	Quantity
Pump: 12 volt DC, 10 amp, demand pump, 60 psi max.	ITT Industries - Flojet	FloJet pump model 2100-732 or similar	1
(Optional) Inlet strainer	ITT Industries - FloJet		1
(Optional) Inlet strainer screening for inclusion into reservoir and as cover for reservoir top as well	McMaster-Carr	See catalog for at least 40 mesh	
Reservoir tank	McMaster-Carr, Aquatic Ecosystems, US Plastics	See catalogs	1
Hose clamps, for 5/8" OD hose, 7/32" to 5/8" clamp ID range, 5/16" or 1/2" band width	McMaster-Carr	5011T141	5 or as needed
Hose clamps, for 3/4" OD hose, 5/16" to 1/2" band width, 7/16" to 25/32" clamp ID range	McMaster-Carr	5011T161	5 or as needed
Tygon braid reinforced high purity tubing 3/8" ID, 5/8" OD, 1/8" wall, clear	McMaster-Carr	5624K12	as needed
Tygon braid reinforced high purity tubing 1/2" ID, 3/4" OD, 1/8" wall, clear	McMaster-Carr		as needed

Electrolyte Reservoir and Pump

Description	Supplier	Part #	Quantity
PVDF single barbed tube fitting reducing coupling for ½" x ⅜" tube ID	McMaster-Carr	53055K138	1
PVC spring-loaded ball-check valve ⅜" barb x ⅜" barb, Buna-N seal	McMaster-Carr	7933K33	1
(Optional) Easy-Grip PVC miniature ball valve straight, ⅜" x ⅜" barb	McMaster-Carr	4757K18	2
Easy-Grip PVC miniature ball valve, 3-way, ⅜" barb x ⅜" barb	McMaster-Carr	4757K58	2
SPDT 35 amp toggle switch	Surplus Center	11 2157	2
Barrier strip for electrical connections or similar according to your design. Crimp-on wire terminals, etc.	McMaster-Carr	see catalog	
Structure, stand or housing for pump, filter and switch/ terminal connections, etc.	per design	per design	

Parts numbers and suppliers have been provided for your convenience; however, suppliers may go out of business and parts numbers may change. All parts listed are available from multiple suppliers.

Pressure Tank and Sight Glass

The purpose of the pressure tank is to maintain pressure balance within the system. It is also a pressure snubber for KOH/distilled water delivery into the electrolyzer. The sight glass provides a visual indication of fill level.

The pressure tank consists of:

- ten stainless steel fittings,
- three pieces of small diameter stainless steel tubing,
- one 3" diameter 12" long stainless steel tube,
- two 3" diameter ½" thick stainless steel disks.

Pressure tank assembly

- Drill five $7/16$" holes in the disks, positioned on the disks as in the diagram on page 116. There are two holes drilled in the disk that will be the tank top, and three holes in the disk that will be the tank bottom.
- Tap each hole for a ¼" NPT thread. Tap every hole on a disk from the same side, as the thread is a tapered fit and all fittings enter from one side only.

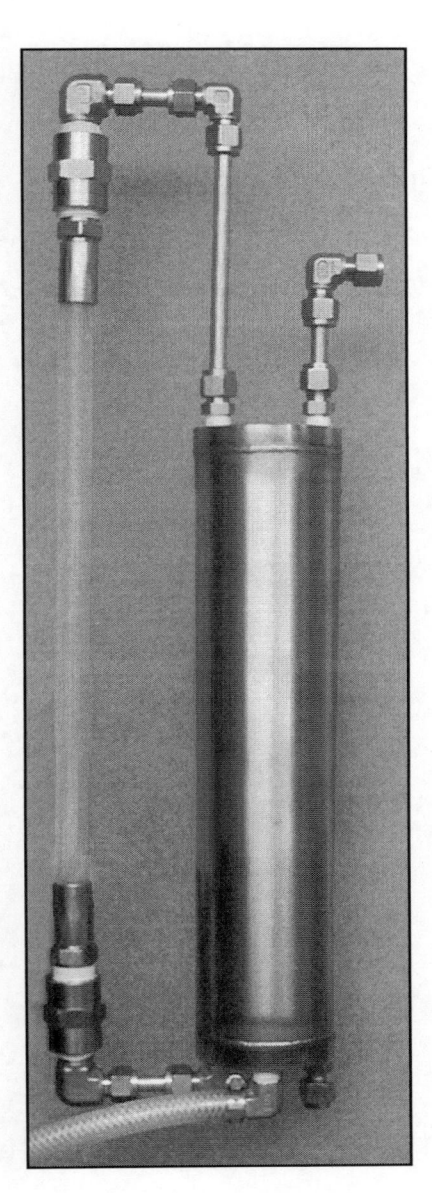

Pressure tank with sight glass

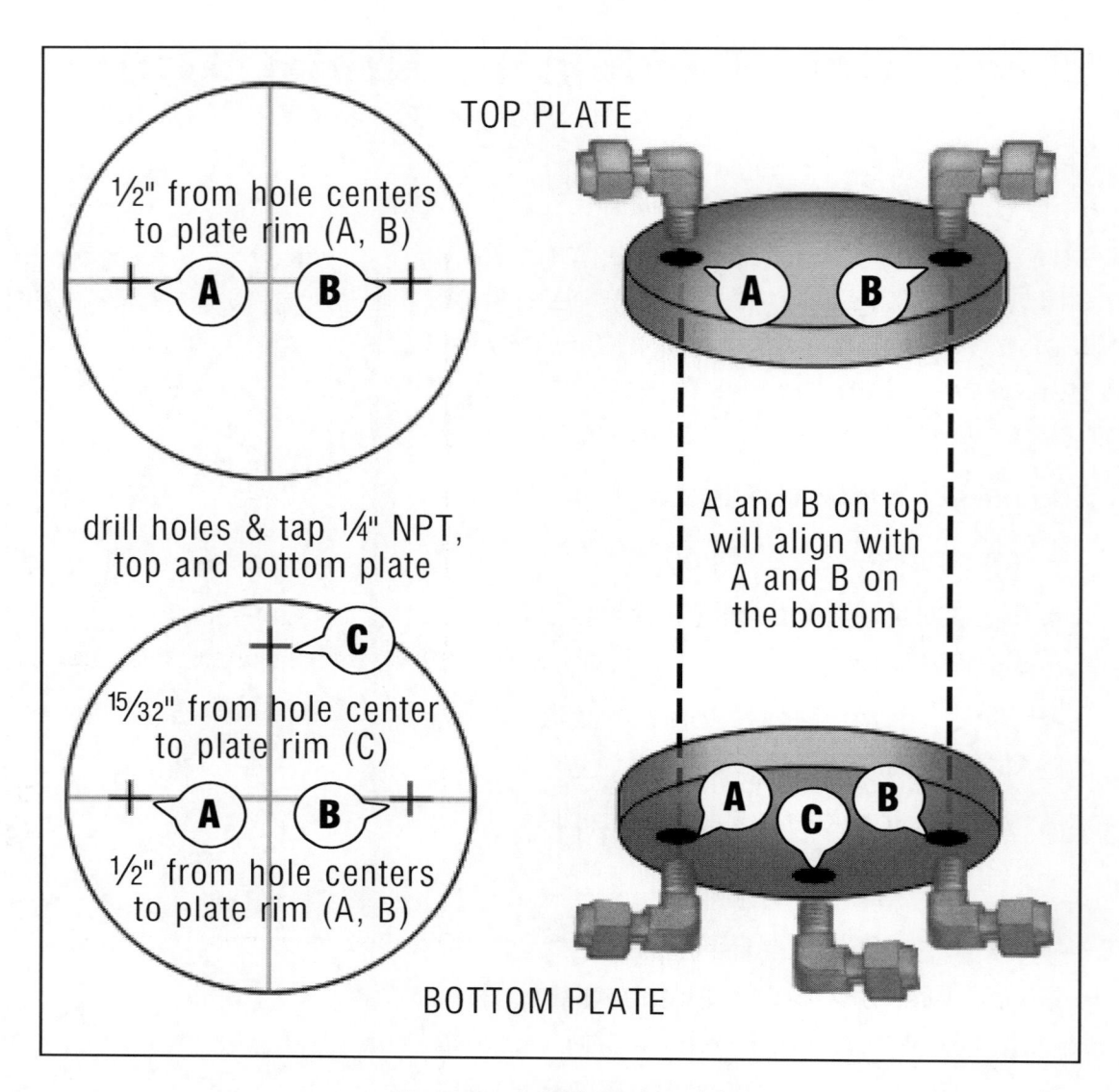

Practical Hydrogen Systems

Pressure tank assembly

Diagram labels:
- TOP PLATE
- ½" from hole centers to plate rim (A, B)
- drill holes & tap ¼" NPT, top and bottom plate
- 15/32" from hole center to plate rim (C)
- ½" from hole centers to plate rim (A, B)
- A and B on top will align with A and B on the bottom
- BOTTOM PLATE

- ◆ Clean and scrub the disks with solvent to remove any cutting oil used and any other grease or grime. While cleaning the disks, also clean and scrub the 3" diameter tube thoroughly in solvent. Rinse these components with water and dry them.
- ◆ Weld the disks to the tube according to the diagram above. Note the relationship of the top disk ports to the bottom disk ports before welding. Make sure that what is to be the outside portion

96

Pressure Tank and Sight Glass

of the top and bottom disks is the surface where the taps begin. If they are welded with the wrong surface inward, you will not be able to screw in the ¼" fitting as it is a tapered fit.

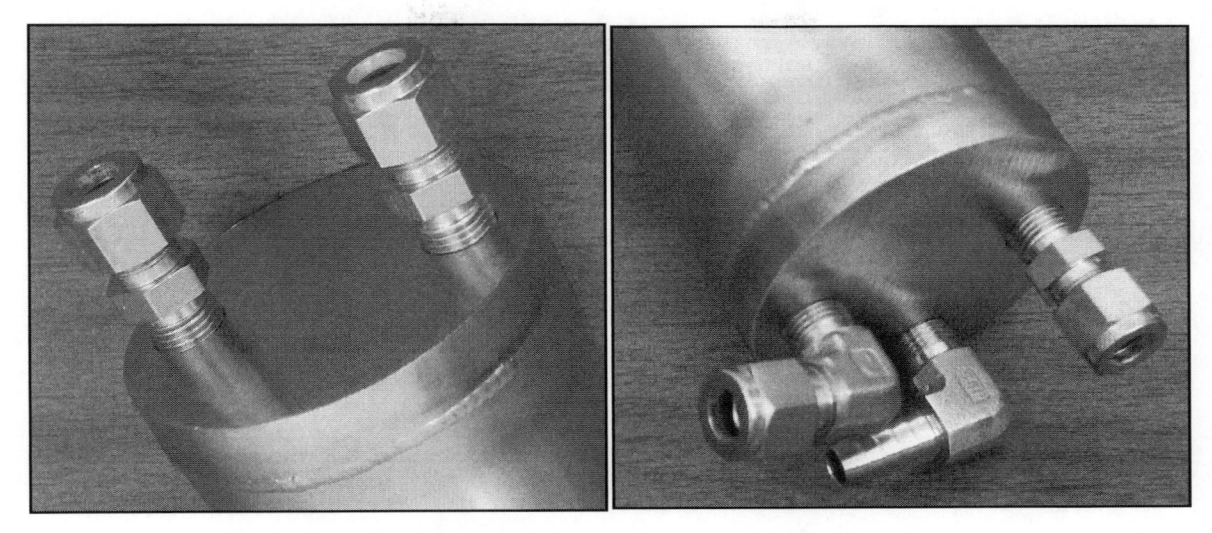

Left, pressure tank top; right, pressure tank bottom

Drilling, tapping and welding can all be done for a reasonable price at a local machine shop if you do not have the time, equipment or skills necessary to complete the job.

After the pressure tank is fabricated, gather the rest of the components to assemble the pressure tank sub-system.

Assemble sight glass and connect to pressure tank

- Using thread tape, sealant or both, attach the two ½" x ½" hex couplers to the sight glass fixture (tubing) on either end. Tighten well.
- Attach the two ½" male NPT x ⅜" tube fitting elbows to the ½" couplings on either end of the sight glass fixture. Tighten them well and be sure that the tube ends are aligned, facing in the same direction.
- Connect the three ¼" NPT x ⅜" tube straight fittings, the ¼" NPT x ⅜" elbow, and the ¼" NPT x ⅜" barb elbow into their appropriate port holes on the pressure tank proper. Make sure that the fittings

are tightened and that the two elbows on the bottom of the tank are pointing in the same direction. Use tape and or thread sealant on all NPT connections.

- Cut three pieces of stainless steel tubing for connecting the sight glass to the pressure tank. Two pieces should be cut to 1⅝" and the third piece should be cut to about 5½". However, before cutting the stainless steel tubing, you may want to purchase some ⅜" diameter wooden dowel rod, and cut that to the suggested tube lengths. Place the dowel pieces into the fittings to see if the lengths make a good fit. The tubing should seat square in the fittings. When you are satisfied that the dowel rod lengths are the correct size, then cut the stainless steel tubing to the appropriate sizes as indicated by the model wooden rod lengths. Make sure to cut each tube square and de-bur with a grinder or small file.

- Clean and scrub the parts to remove any grease and grime.

- Place the tubing into the tubing ends of the compression fittings and tighten. Do not use any tape or compound on the threaded nuts that compress the tube end. Tape or other materials on the threads on the tube end will interfere with proper tightening and seating of the compression rings.

This completes the pressure tank sub-system.

Parts numbers and suppliers are provided for your convenience; however, suppliers may go out of business and parts numbers may change. All parts listed are available from multiple suppliers.

Pressure Tank and Sight Glass

Pressure tanks and sight glass parts

Description	Supplier	Part #	Quantity
High-polish stainless steel tubing type 316/316L, 3" OD, 2.87" ID, .065" wall, 1' L	McMaster-Carr	4466K491	1
Type 316 stainless steel disc 3" diameter, ½" thick	McMaster-Carr	9260K51	2
Full-view flow and level sight ½" NPT male, 12" sight length, 17" overall length	McMaster-Carr	5071K31	1
Type 316 stainless steel hose adapter 90° elbow, barbed x male, ⅜" hose ID, ¼" pipe	McMaster-Carr	53505K43	1
316 SS Yor-Lok compression tube fitting 90° elbow, tube x male for ⅜" tube OD, ½" NPT	McMaster-Carr	5182K158	2
316 SS Yor-Lok compression tube fitting male straight adapter for ⅜" tube OD, ¼" NPT	McMaster-Carr	5182K119	3
316 SS Yor-Lok compression tube fitting 90° elbow for ⅜" tube OD	McMaster-Carr	5182K416	1
316 SS Yor-Lok compression tube fitting 90° elbow, tube x male for ⅜" tube OD, ¼" NPT	McMaster-Carr	5182K156	1
Hex coupling NPT ½" female x ½" female, stainless steel	McMaster Carr	48805K79	2
Type 316 stainless steel seamless tubing ⅜" OD, .305" ID, .035" wall	McMaster-Carr	89795K837 (see also p.206)	one 6' length

Electrolyzer and Electrolyte

The purpose of the alkaline electrolyzer in this system is to disassociate water into hydrogen and oxygen. This is accomplished by applying low voltage, high amperage DC current through two electrodes. The electrodes are located inside the electrolyzer and immersed in a solution of KOH (potassium hydroxide). The KOH improves the conductivity of the water and accommodates the process of disassociation by providing a conductive path for ionic movement within the water.

How the electrolyte works

Potassium hydroxide (also known as caustic potash) is a strong electrolyte. This means that it is essentially 100% ionized in solution and thus is a good conductor of electricity. When the positive and negative poles of the electrolyzer are connected to the power source, hydrogen ions combine with electrons at the negative electrode to form hydrogen, and hydroxy ions give up electrons at the positive electrode, releasing oxygen.

Twice as much hydrogen is generated as oxygen, since a water molecule contains two hydrogen atoms for every oxygen atom. Other electrolytes, such as sodium hydroxide, can be used, but they are not as conductive as KOH. An acid such as sulfuric acid can be used as an electrolyte, but it is more corrosive to the electrodes. The added wear and tear on the electrodes and other components does not justify its use. Essentially, KOH is the best choice for alkaline electrolyzers.

Obtaining potassium hydroxide

KOH can be purchased, or you can make it yourself if you have a source of hard wood ashes. The making of lye used to be a common chore in most households. If you are interested in this, my book *Build Your Own Fuel Cells* contains complete illustrated instructions for making KOH from wood ash.

Practical Hydrogen Systems

Whether you purchase KOH from a chemical supply house, or make it yourself, the following table will be helpful. If you make KOH yourself, use the table to determine the specific gravity of the solution when you take a hydrometer reading. This will indicate whether to boil the solution down to strengthen it, or add more distilled water to weaken it.

Potassium hydroxide (KOH) solution strength

Specific Gravity	Percent KOH	Lbs. per Gallon	Specific Gravity	Percent KOH	Lbs. per Gallon
1.0083	1	0.0841	1.1493	16	1.535
1.0175	2	0.1698	1.159	17	1.644
1.0267	3	0.257	1.1688	18	1.756
1.0359	4	0.3458	1.1786	19	1.869
1.0452	5	0.4361	1.1884	20	1.983
1.0544	6	0.528	1.1984	21	2.1
1.0637	7	0.6214	1.2083	22	2.218
1.073	8	0.7164	1.2184	23	2.339
1.0824	9	0.813	1.2285	24	2.461
1.0918	10	0.9111	1.2387	25	2.584
1.1013	11	1.011	1.2489	26	2.71
1.1108	12	1.112	1.2592	27	2.837
1.1203	13	1.215	1.2695	28	2.966
1.1299	14	1.32	1.28	29	3.098
1.1396	15	1.427	1.2905	30	3.231

Managing electrolyte strength

We make our own KOH and do not bother to boil it down. Initially, KOH solution is put in the electrolyte reservoir. The KOH, when it is first made, is at about a 10 to 12% solution. In the electrolysis process, the KOH solution becomes stronger as the water disassociates and is used up, leaving the KOH behind. Since more KOH solution is

Electrolyzer and Electrolyte

added to the system to maintain the proper level instead of distilled water, the solution gets stronger over time. Once the solution in the electrolyzer is at the ideal specific gravity, the electrolyte reservoir is replenished with distilled water instead of KOH solution.

If you start with the exact specific gravity that you want when you first fill the reservoir, then afterward, simply replenish the reservoir with distilled water to the same level of your first fill. KOH is lost over time, but it is mostly the water that is used up in the process.

Handling KOH

- Wear your protective gear!!!
- Be sure to use distilled water only. Well, tap, and spring water contain far too many unknowns (minerals, organic particles, etc.) that will cause problems with the electrolyzing process and gum up the electrodes.
- Never mix dry purchased KOH into the water in the reservoir – the process generates too much heat. Always mix KOH into the water by putting a little bit in at a time, very slowly. Do not mix by pouring water on to the KOH.
- Use only plastic or stainless steel buckets or containers. Do not use aluminum pots or utensils for mixing or holding KOH solution!!!
- And, always let the solution cool down before refilling the reservoir.

For best conductivity, the solution should be about 29.4%, which means it would have a specific gravity of about 1.28. This would require 3.231 pounds of KOH for one gallon of distilled water. This particular specific gravity is not necessary, and a milder, less conductive solution of about 12% will work. For optimal performance however, a 29.4% solution is the best.

To check the specific gravity of a solution, use a hydrometer such as those designed for testing the specific gravity of battery electrolyte. Hydrometers can be purchased at any auto parts or hardware store.

Using KOH safely

Never forget that KOH is extremely caustic and can cause severe burns and blindness if not handled properly. Please do not ignore these warnings. Make sure you have eye protection such as safety glasses or a safety face shield. These can be purchased inexpensively at most hardware stores. Always cover all your skin completely with protective clothing and rubber gloves when working with and around KOH; and study and follow all MSDS recommendations. MSDSs are available online.

Be diligent with all safety precautions when handling KOH, and keep it out of reach of children and pets – this is a very dangerous material! If you are operating an experimental hydrogen system with KOH under pressure, you should be aware that a leak in the system can mean that the caustic KOH is sprayed out. Never allow humans without safety garb, or animals, in the vicinity of an operating experimental system without appropriate shielding.

Ready to go!

Electrolyzer construction

The body of the electrolyzer consists of a 10" length, 3" NPT schedule 40 stainless steel pipe nipple and four stainless steel flanges.

Attendant components are six ¼"NPT x ⅜" tube, straight, compression type fittings which serve as electrode ports and gas and KOH/water distribution ports.

Electrolyzer and Electrolyte

Screw-on flanges. Left, gasket side; right, pipe nipple side

Flanges

Two flanges form the top, and two form the bottom of the electrolyzer. These pairs of flanges consist of one screw-on flange and one blind flange.

The screw-on flanges, as the name implies, are screwed on to the body of the electrolyzer (pipe nipple). The blind flanges are bolted to the screw-on flanges, with a Teflon gasket placed in between each blind and screw-on flange. When the bolts and nuts are tightened, the pair of flanges are sealed.

Using the screw-on flanges gives some flexibility to experiment with different electrolyzer designs, such as pipe nipple caps for lids, etc. The disadvantages of the screw-on flanges and threaded pipe

Blind flanges. Left, outside; right, gasket side

nipple is a lower pressure rating for the vessel, and leak problems. The threads on this nipple are very large and it is not as easy to fill and close the gaps as with smaller fittings.

Alternative designs

A pipe nipple without threads and with a weld-on flange would be a more permanent fixture. Also, the screw-on flanges can be welded on, though it would be best to use weld-on flanges if you are going that route. Weld-on flanges will reduce leak problems, and the pipe nipple wall structure would not be weakened by threading.

Another alternative is to epoxy the surrounding union point of the screw-on flange after applying tape and sealant and screwing it on. Of course, if the components are welded or sealed with epoxy, you will not be able to open up the electrolyzer and make design changes.

It's not that a firm, leak-free joint cannot be made, but it does take some muscle power to twist the screw-on flange tight enough so that minor leaks do not occur. We sealed and taped the nipple, and then screwed on the flange as tightly as possible by hand. We used two people – one person to hold one flange and one person hold the other flange, both twisting as hard as they could in opposite directions. This method worked well. We did not want to tighten the flanges to the point where they would be extremely hard to remove, as we wanted to maintain a measure of flexibility in the system.

Electrolyzer assembly

Flange ports

- Mark the centers for the four holes on the blind flanges, first, the two at dead center of each blind flange. Then, for the top flange oxygen outlet, mark a point 1 9/32" from the flange center. For the bottom flange electrolyte port, mark a point 1 1/16" from the flange center.
- Drill 7/16" holes in the two blind flanges as marked, two holes in each flange.

Electrolyzer and Electrolyte

Machining a flange

◆ Tap the four holes just drilled in the blind flanges for ¼" NPT thread. Do not tap the thread all the way through the flanges. Tap each flange from the flat side, which will be the outside when electrolyzer is constructed. The flange faces that will be inside have a raised surface for gasket placement.

In the top flange, the off-center hole where the oxygen outlet will be is a very close fit (1⁹/₃₂" center to center) when the PVC coupler is installed. This hole could be moved very slightly towards the circumference, but then the inner wall of the pipe nipple will interfere ever so slightly with the oxygen port. Our choice of placement had to do with using mainly off-the shelf parts, and this placement suited our purposes quite well.

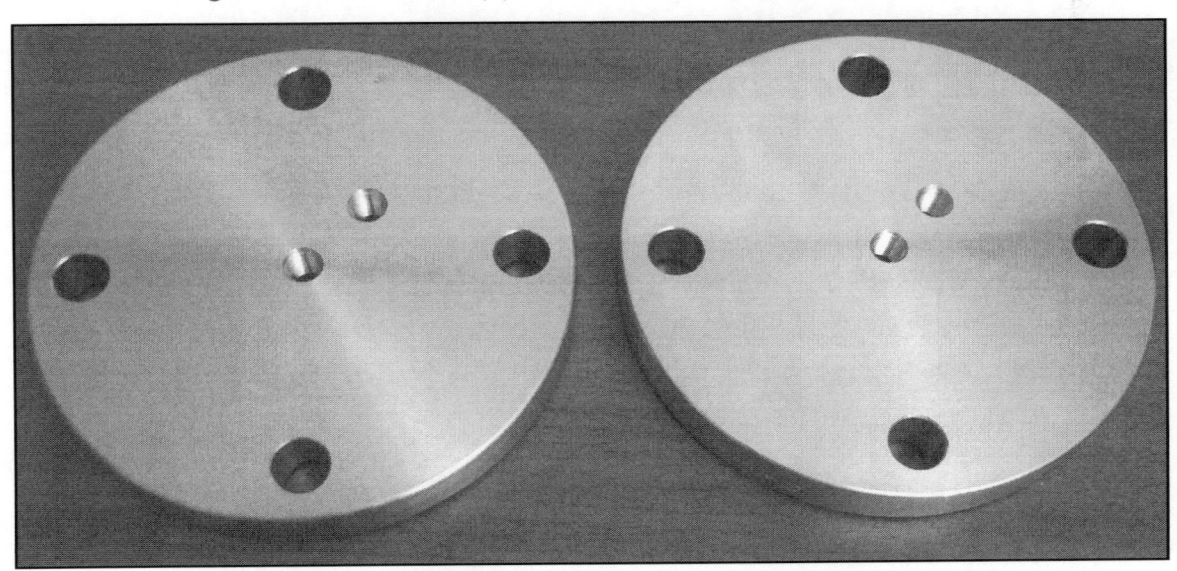

Blind flanges, drilled and tapped. Left, top flange; right, bottom flange

One option would be to use a welded-on stainless steel duct with thinner walls welded to the inside of the flange instead of the thicker walled PVC coupling, or find a PVC coupling with thinner walls. We chose the thick walled PVC coupler because the wall thickness gave us more glue surface on the edge of the coupler to act as a barrier for the gases. This does not mean that another approach, or different dimensions will not work as well.

Pipe nipple for electrolyzer body, drilled and tapped.

Positive electrode ports

- Drill two 7/16" holes 180° apart in the pipe nipple. Do not drill a hole on the pipe weld seam. The hole centers should both be 6" from the same end of the pipe nipple.
- Tap both holes for ¼" NPT thread. Do not tap too far. The fittings for the positive electrodes should only extend slightly past the inner surface of the pipe wall, so that they do not interfere with the liner that will be installed later.
- Clean and scrub the flanges and pipe nipple with solvent to remove oils and grime.

Some of these parts, such as the threads on the pipe nipple, may have burrs on them, or shavings of metal that were not completely removed in the threading process done by the manufacturer. Pull these off, or grind them, or file them down. Inspect all parts as you receive them and finish working on them, and remove any such imperfections.

Electrolyzer and Electrolyte

Preparing the Teflon liner

A Teflon sheet is used as a liner for the inside of the electrolyzer. It prevents contact between the metal mesh positive electrode and the metal inner walls of the electrolyzer. The Teflon that we used has one adhesive-compatible side which is brown. This adhesive-compatible side faces inward when constructing the electrolyzer so that spacer strips of silicone rubber can be glued to its surface. The white side of the liner is the outside of the liner and faces the inner wall of the pipe nipple.

Teflon liner

- Cut a 10" x 9⅝" rectangle from the Teflon sheet.

- Roll the liner into a cylinder with the 9⅝" length for the circumference and the 10" for the length of the cylinder. The brown (adhesive compatible) side should be the inside of the cylinder.

- Slip the Teflon liner into the nipple and note if there is any overlap. The liner lies against the inner wall of the pipe nipple, and the ends of the liner should abut but not overlap. If there is overlap, trim it away carefully until the edges fit properly. Due to possible irregularities in the pipe nipple, it is wise to cut the liner to a give a generous circumference for the liner (9⅝" is generous). It is better to have to trim it to size, than to ruin the piece by having cut it too short.

Fitting the Teflon liner

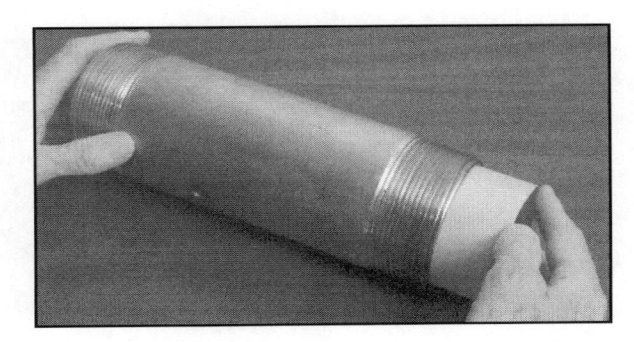

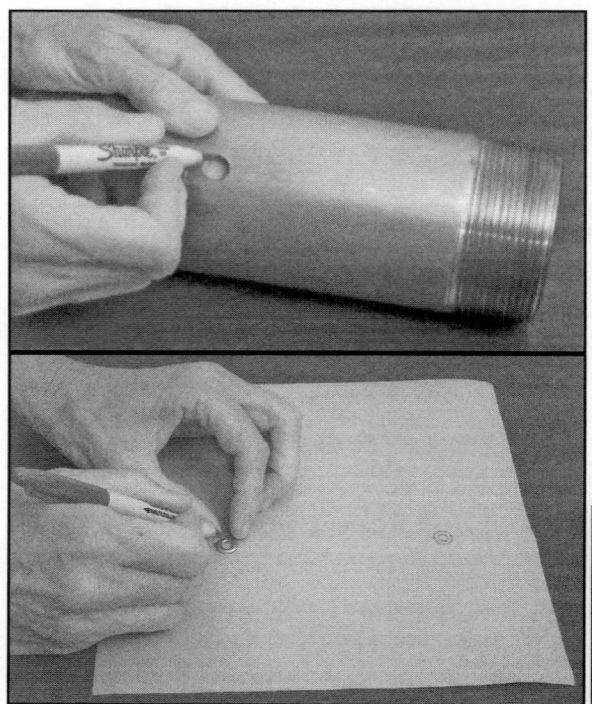

Practical Hydrogen Systems

♦ After the liner is trimmed, slip it into the nipple. The position of the liner's seam when placed in the nipple should be 90° from each positive electrode port hole in the nipple (halfway in between them). Mark the position of the holes through the pipe nipple.

Mark the port holes for the positive electrode assembly on the liner.

♦ Take the liner out of the nipple and punch the holes out as marked, using a $7/32$" punch.

Punch the port holes in the Teflon liner.

♦ Cut four silicone rubber strips from a $3/32$" thick sheet. These strips should be $3/8$" wide and 10" long.

♦ Mark four lines on the brown side of the liner that will go along the length of the cylinder. Starting from the left edge, the lines should be at about 1¼", 3½", 6" and 8¼" from the edge.

Electrolyzer and Electrolyte

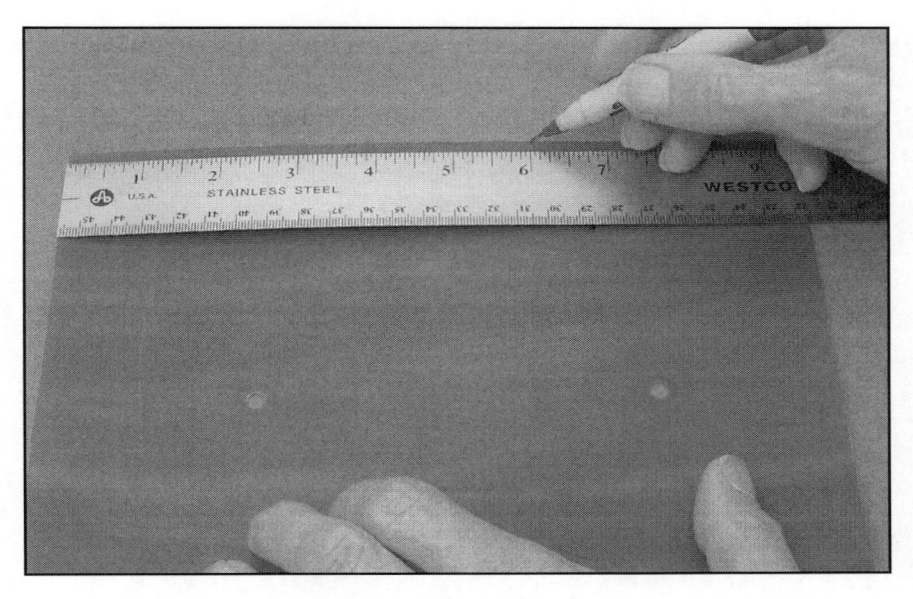

Mark the positions for rubber spacer strips

- Lightly sand one surface of each silicone strip. This will aid in adhesion. Make sure to clean the surface of the rubber strips after abrasion.

- Apply epoxy to the sanded surface of each silicone rubber strip and lay each one on the liner, centered along the lines drawn. Let this dry for 24 hours.

Teflon liner with spacers glued on

Practical Hydrogen Systems

- Attach the top and bottom screw-on flanges to the pipe nipple.
- First, test by screwing on the flanges to the pipe nipple to see generally how much travel you get. The screw-on flanges will not cover all the threads on the nipple. This test will indicate how much of the nipple to thread seal and tape.
- Thread seal and Teflon tape each end of the nipple, covering the surface indicated by the test.

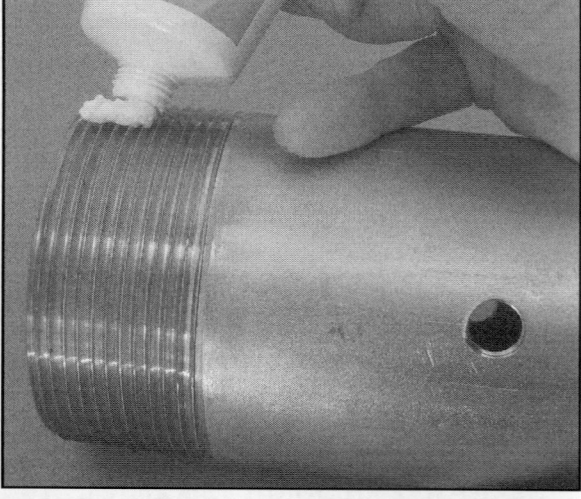

- Screw on the flange. It is helpful to have two people to do this, each holding one of the flanges and twisting in the opposite direction. Other options are to use a special jig or vise to hold the nipple while you tighten each flange.

Electrolyzer and Electrolyte

Electrolyzer top

- Lightly sand the areas where the PVC pipe coupler will be in contact with the surface of the inside of the top flange. Sand the contact surfaces on both the flange and the coupler.

- Remove all sanding dust by washing before you epoxy. This will help adhesion.

- Epoxy the pipe coupler hydrogen duct to the top blind flange. This must be centered on the center hole of the flange. Epoxy very thoroughly so that no gas can leak into or out of this coupler in operation.

The coupler, along with the felt separator it holds, prevents the mixing of hydrogen and oxygen inside the electrolyzer, which can be dangerous. Hydrogen should be inside the coupler, and oxygen outside the coupler. Notice that the outer edge of the coupler abuts the oxygen port hole very closely, or may even hang over it very slightly. This is OK because the wall thickness of the pipe coupler is sufficient that the glue surface will keep the gases separate, if it is glued well. It's also wise to coat the inner and outer circumference of the coupler where it contacts the flange to ensure a better seal.

Positive electrode assembly

The positive electrode assembly consists of two electrode feed-throughs, and a circular mesh electrode.

Each positive electrode feed-through consists of:

- two rubber washers
- one nickel washer
- one steel fender washer
- one compression fitting
- one 8-32 threaded rod
- five 8-32 nuts
- one drilled and threaded Teflon insert
- Teflon tape and thread sealant.

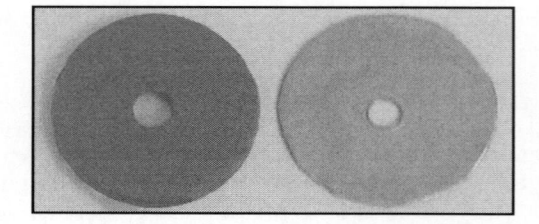

Rubber washer and nickel washer

Make the silicone rubber washers

- Lay a 1" fender washer on a piece of $3/32$" silicone rubber sheet and trace the outside diameter of the washer onto the rubber sheet.
- Cut the outside diameter with shears, scissors, or a razor knife.
- Punch or drill a $1/8$" center hole. The center hole only has to be large enough to accommodate the threaded rod.

Each electrode feed-through has two of these silicone rubber washers, for a total of four in the positive electrode assembly.

Make the nickel washers

- Lay a 1" fender washer onto a nickel sheet and trace the outside diameter of the washer onto the nickel.
- Cut the washer out of the sheet with a metal shear
- Punch or drill the center hole out $1/8$".

Each electrode feed-through has one nickel washer, for a total of two in the positive electrode assembly.

Electrolyzer and Electrolyte

Cut the electrode rods

◆ Cut the 8-32 threaded stainless steel rod to two 2⅝" lengths. Each positive feed through needs one, so total, two.

Preparing the Teflon insulators

One threaded Teflon insert for each feed-through (total two in the positive electrode assembly).

◆ Cut two ⅝" long pieces from a ⅜" diameter Teflon rod.

◆ Drill a hole through the center of the ⅝" Teflon rods with a #29 drill.

Mark the pieces of Teflon rod

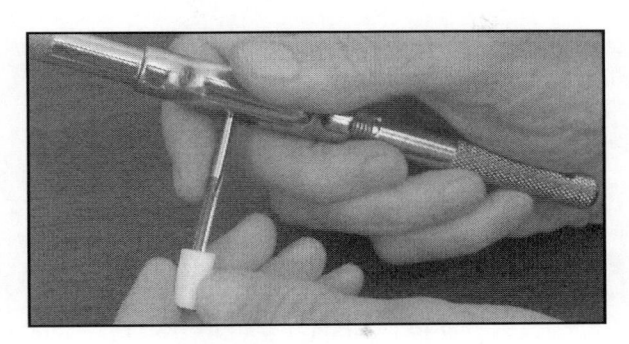

Tap the pieces of Teflon rod

◆ Tap threads with a 8-32 hand tap.

The other components needed for the positive electrode feed-throughs are two 1" fender washers, ten 8-32 stainless steel nuts for each feed-through, and the two compression fittings.

Completed Teflon insulators

Practical Hydrogen Systems

Fitting the positive electrode feed throughs

- Screw the threaded rod into the Teflon insert.
- Put the ferrules on the Teflon (see photo above), insert into the tubing end of fitting and tighten lock nut slightly (see page 68).
- Screw the rod until the rod extends out from the ¼" NPT side by 3/16" or slightly more.
- Take the nut off and pull the threaded insert out of the fitting.
- Apply thread seal and Teflon tape from the end of the insert to 3/16" from end. Wrap the Teflon tape around the rod at east three times. The Teflon tape is an electrical insulator for the rod and keeps it from inadvertently touching the walls of the compression fitting at its exit point.
- Back off the Teflon insert by unscrewing it ⅝".

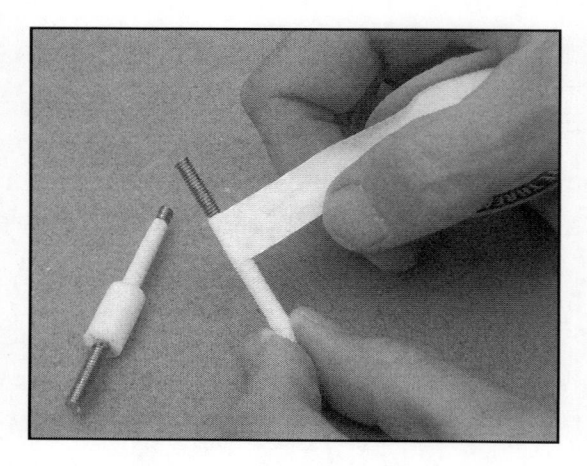

Teflon tape the positive electrode rod

Electrolyzer and Electrolyte

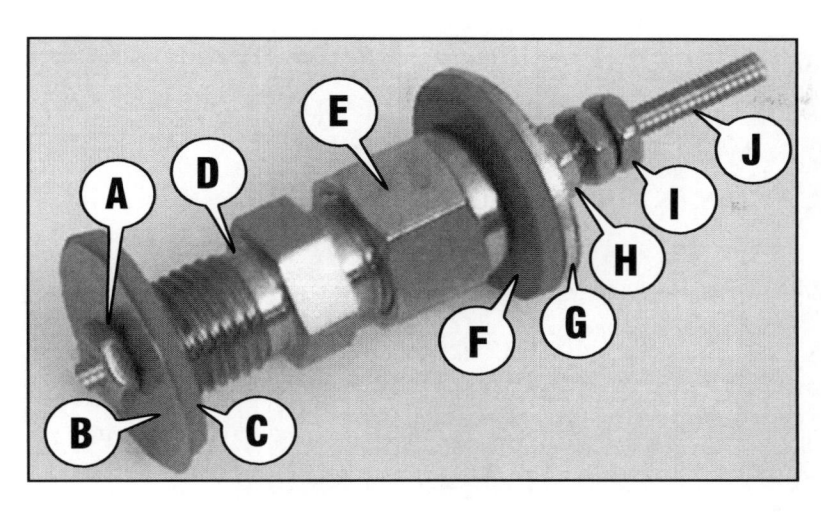

A	Nut
B	Nickel washer
C	Rubber washer
D	Fitting body
E	Fitting nut
F	Rubber washer
G	Stainless steel washer
H	Pair of nuts
I	Pair of nuts for connector
J	Threaded rod

◆ Apply thread seal to the bare ⅝" of the rod you have just exposed

◆ Screw the Teflon insert back on over the part of the rod you applied the thread seal to. Do not use too much thread seal – a little will suffice.

Positive mesh electrode

◆ Cut a rectangle 9⅞" x 7½" from the monel woven wire cloth (mesh) for the positive electrode mesh. This allows an overlap of about ¼" so that the two surfaces can be tabbed together with nickel foil tabs.

◆ Insert the Teflon liner about ¾ of the way into the nipple. Make sure the liner edges are abutting and even.

Practical Hydrogen Systems

- Roll up the mesh and slip the screen into the liner. The mesh must touch the rubber spacers, but there should be space between the screen and the liner where there are no rubber strips. When the mesh is positioned correctly, it should overlap itself about ¼".

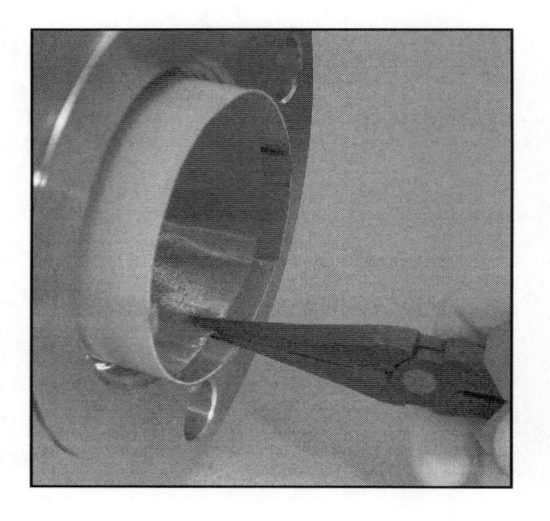

- With needle nose pliers, pull out the mesh at the overlap and pinch a nickel tab onto the overlap, binding the mesh edges together.

- Take the mesh out, align it and put a nickel tab on the other end. Both ends of the mesh are now tabbed.

- Slip the mesh back into the liner and position it about ¼" from the top of the liner.

- Pull the liner out until the liner holes are visible. Mark both of these holes on the mesh.

Electrolyzer and Electrolyte

- Pull out the mesh and with a sharp object, separate the mesh enough so that the 8-32 threaded rod will pass through it

- Remove the Teflon liner from the pipe nipple.

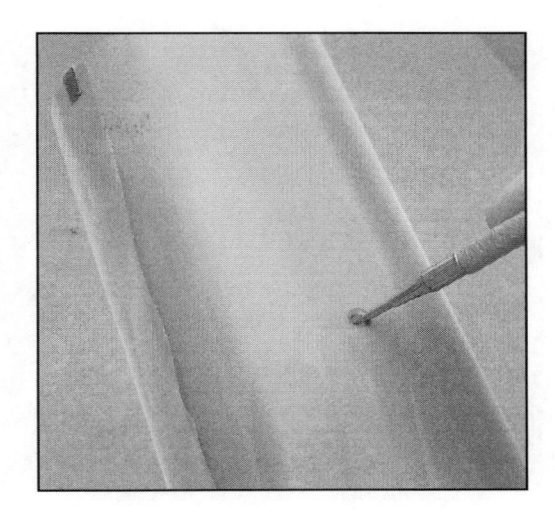

Positive electrode installation

- Coat two ¼" NPT x ⅜" Yor-Lok fittings on the ¼" NPT side with thread seal, and tape with Teflon. Screw these into the positive electrode port holes and tighten.

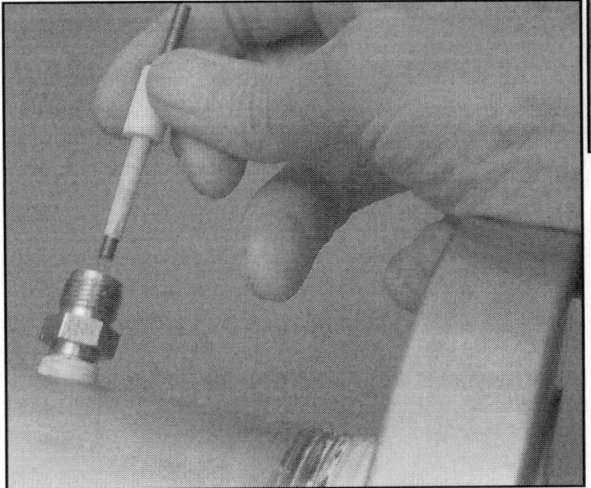

- Place the taped rod with the Teflon insert into the fitting, put on the ferrules and nut, and then tighten.

Tighten the ferrules and nut on the taped rod.

- Insert and position the Teflon liner back into the nipple. The threaded rod will now protrude through the holes in the liner when it is positioned correctly.
- Place the silicone washers on the protruding rods.

Electrolyzer and Electrolyte

- Insert the mesh, position and align, slipping the mesh over the protruding electrode rods. Two barbecue sticks can be used to help guide the mesh so that it does not catch on the threaded rod before it is correctly positioned.

- Line up the holes in the mesh with the electrode rods and push the mesh gently on to the rods.

- Put a nickel washer on the inside of each electrode rod, over the mesh.

Practical Hydrogen Systems

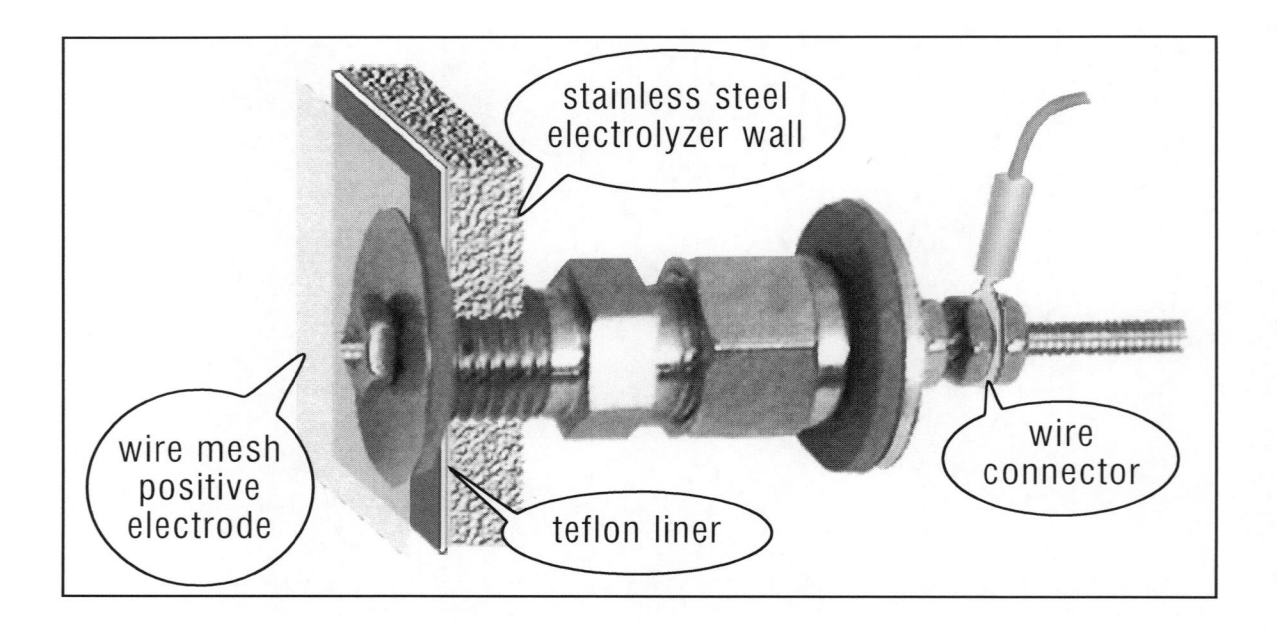

labels: stainless steel electrolyzer wall; wire mesh positive electrode; teflon liner; wire connector

◆ Screw the 8-32 nuts onto the rod over the nickel washers to fasten and tighten.

Inserting the flange fittings

Apply thread seal and Teflon tape to the NPT ends of the straight ¼" NPT x ⅜" tubing Yor-Lok fittings, and screw and tighten them into the four holes in the top and bottom blind flanges.

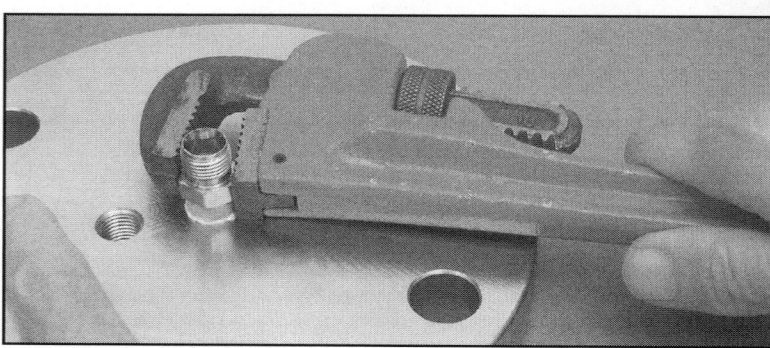

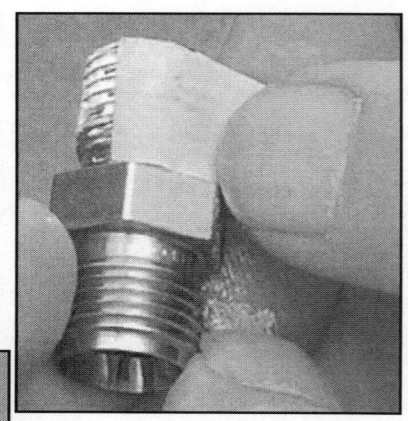

Electrolyzer and Electrolyte

Negative electrode preparation

- Cut a 12" long piece of 8-32 threaded rod.

- Cut a 5/8" long piece of 3/8" diameter Teflon rod to serve as an insert.

- Drill a hole in the center of the 3/8" diameter Teflon rod with a #29 drill and thread with a 8-32 tap.

- Drill a hole in the center of the 3/4" PVC pipe cap and thread with an 8-32 tap.

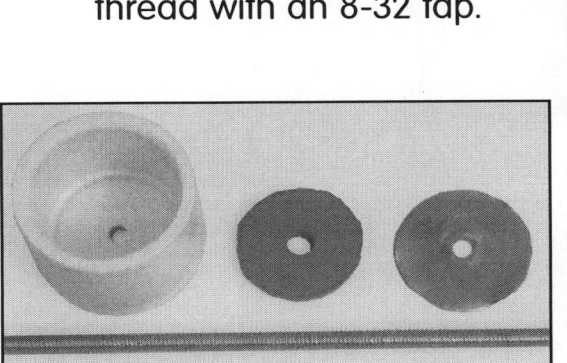

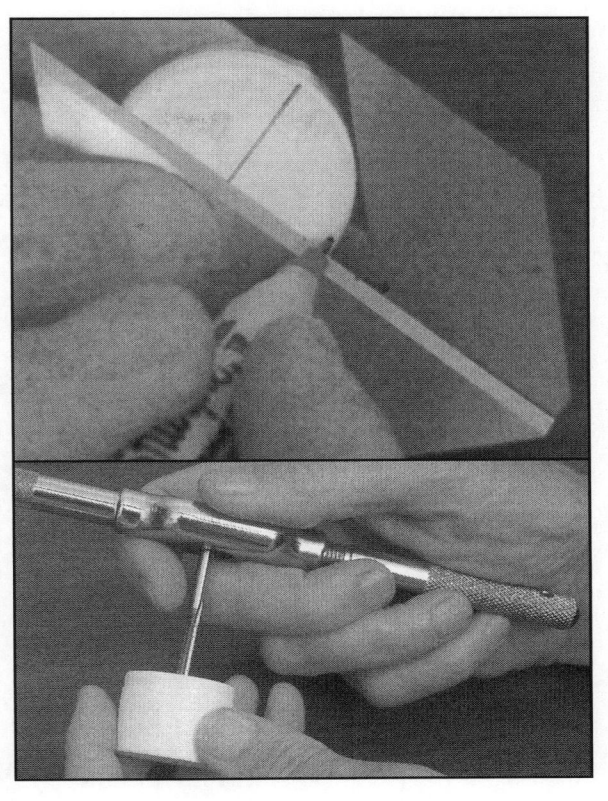

- Cut two 1" diameter silicone rubber washers and one 1" diameter nickel washer. Drill holes in center to accommodate 8-32 threaded rod.

- Cut a 3/8" x 3/8" square tab from nickel foil. Drill a hole in the center to accommodate 8-32 rod.

- Cut two 1/8" wide strips, 1/2" to 3/4" long, of nickel foil for tabs to crimp together the edges of the pleated negative electrode, one at the top, one at the bottom. Bend them partially in half across the 1/8" width, to prepare them.

- Cut a 3/8" diameter nickel stop tab with two 1/8" wide and 1/2" long tabs on each end. Drill a center hole in the stop tab to accommodate 8-32 rod.

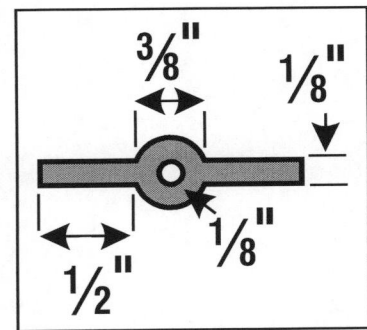

123

Negative electrode mesh

The negative electrode mesh is pleated and rolled into a cylinder.

- ◆ Cut the monel wire mesh for the negative electrode as shown in the illustration. The main rectangle is 10⅛" x 7½", and the tab is ¾" wide and 7½" long. The tab begins 1⅞" from the edge of the large rectangle.

- ◆ Mark the fold points for the pleats on each side of the mesh. The fold points should be ⅜" apart.

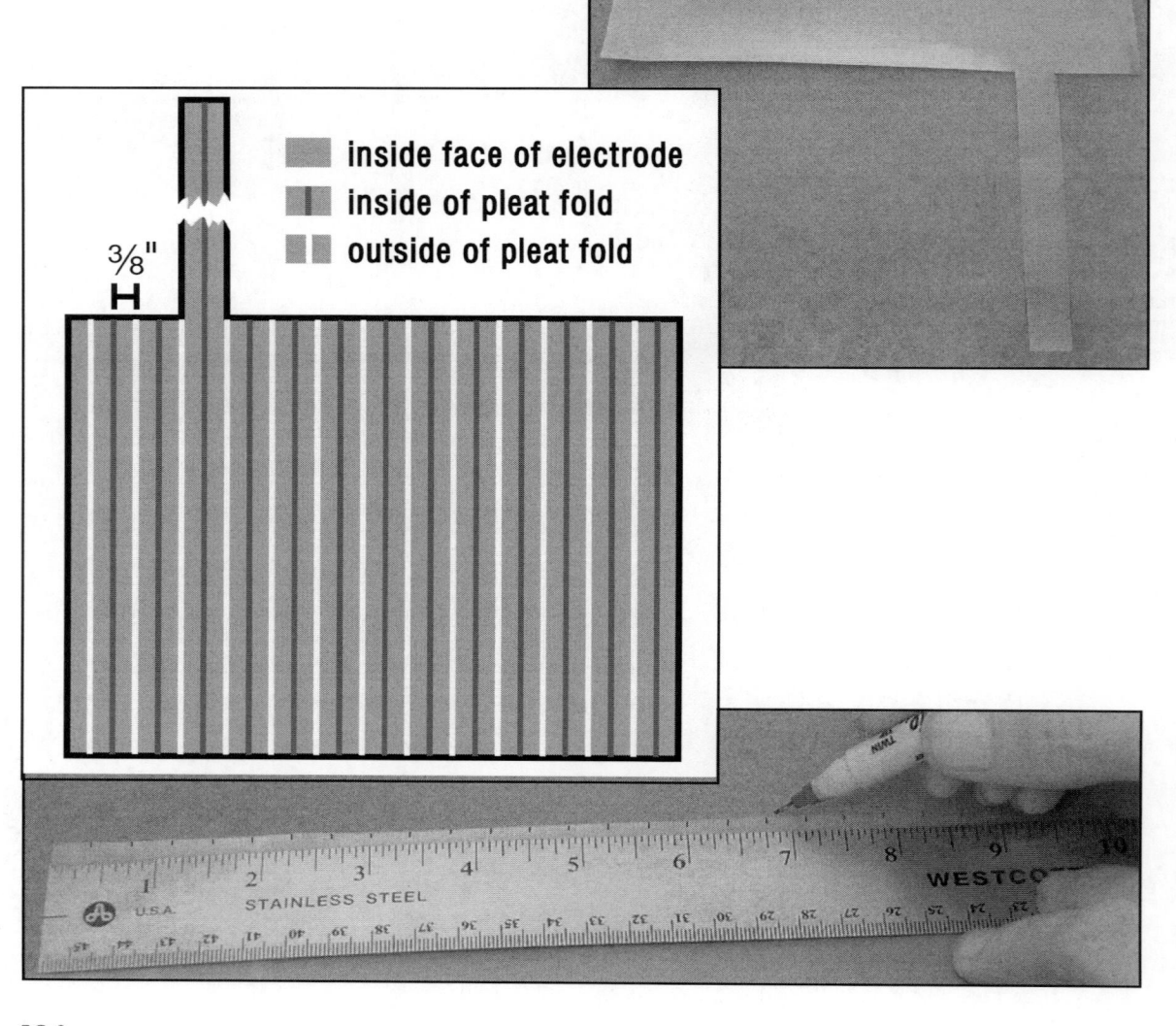

Electrolyzer and Electrolyte

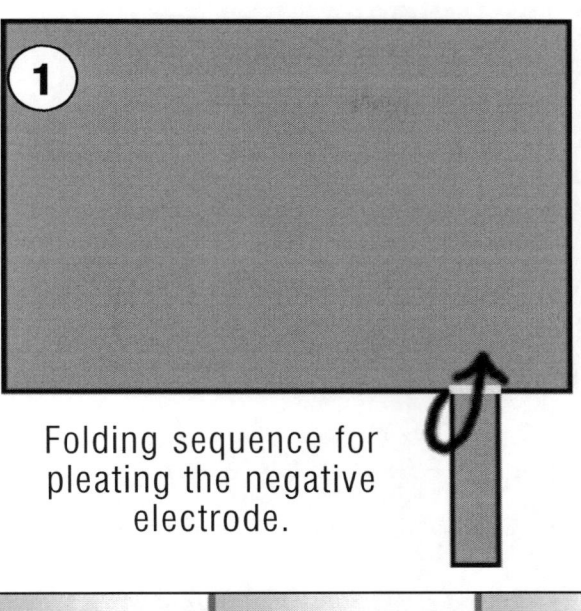

◆ Fold the mesh as shown in the illustrations, using a stainless steel ruler to make the creases of the folds sharp. (See photos below.)

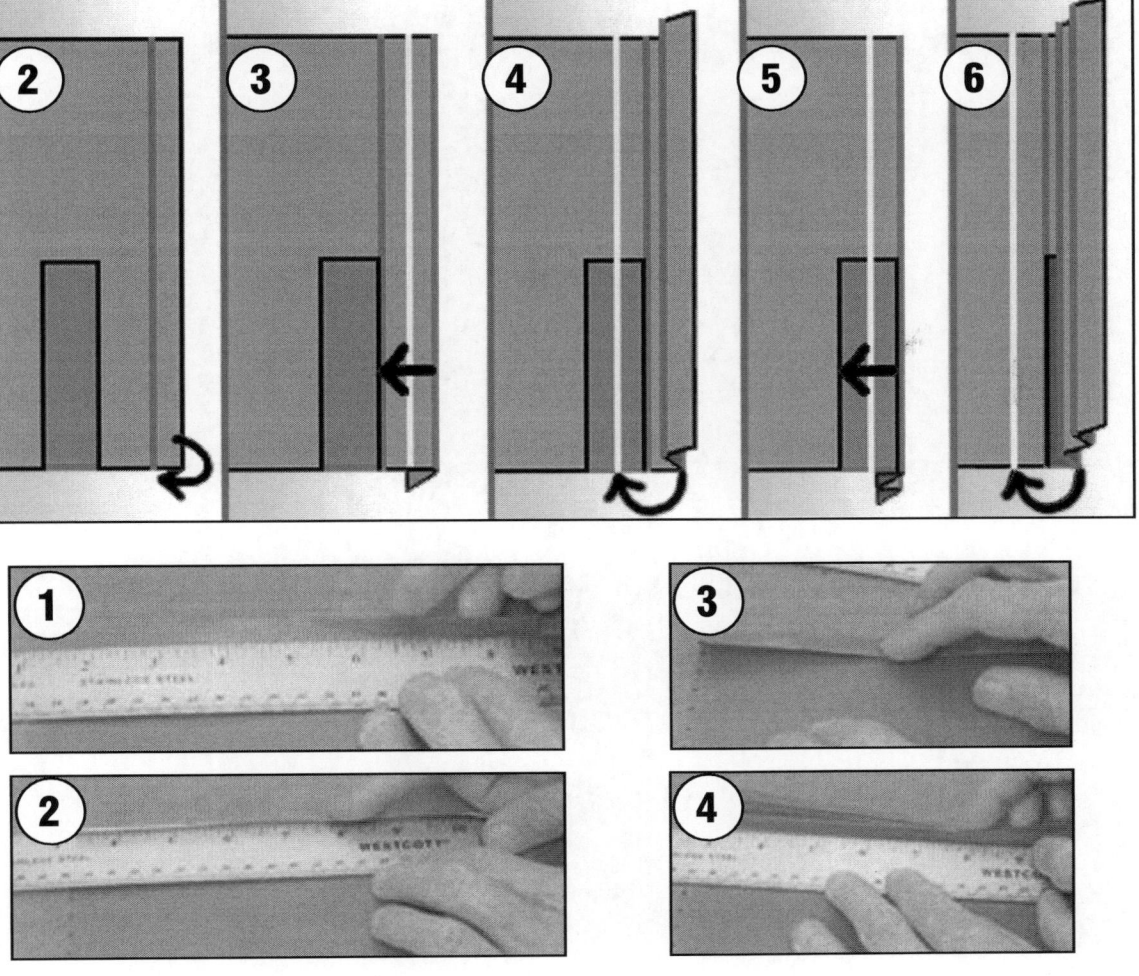

Folding sequence for pleating the negative electrode.

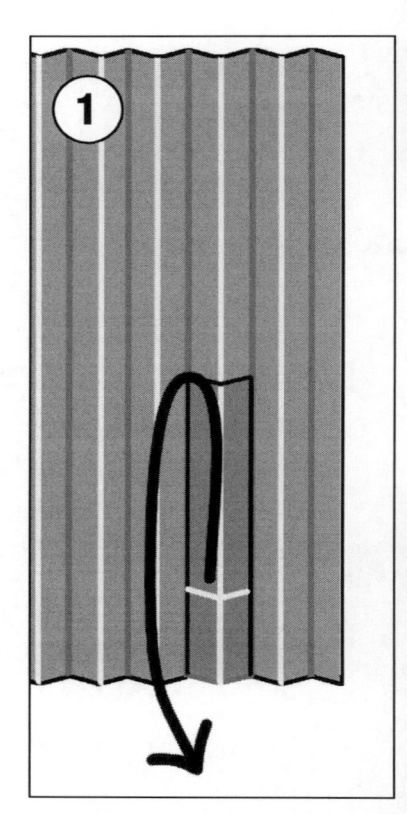

Practical Hydrogen Systems

♦ When all the pleats have been made, fold the tab back along the edge of the main piece towards the pleats and crease the fold, then finish the tab by folding as shown. Punch a hole for the screw through all layers of the tab.

Folding sequence for negative electrode tab connector

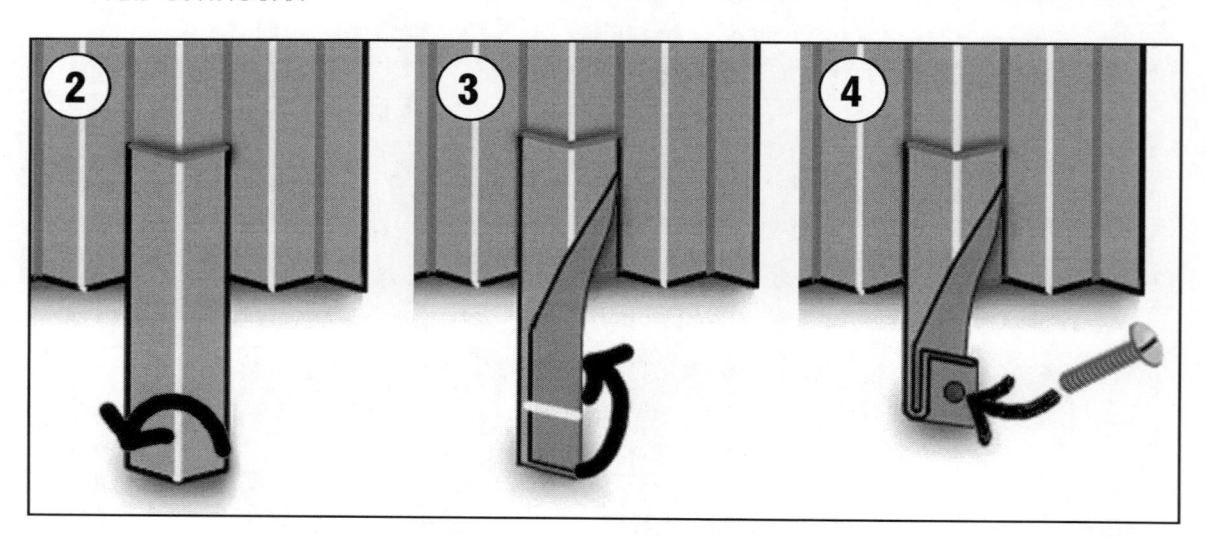

Electrolyzer and Electrolyte

- Roll the pleated mesh into a cylinder with the tab coming out from the inside of the roll, and overlap the edges as shown.

- Crimp the overlapped mesh together with the nickel tabs.

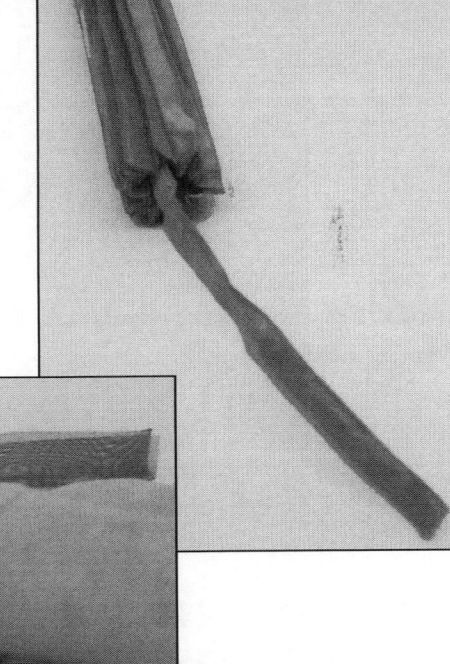

127

Practical Hydrogen Systems

- Fold the end of the tab extension back on itself. Leave just enough length so that when you place the tab extension in the PVC cap, there is enough room to reach in between the bottom of the pleats and the PVC cap, and screw the nut down firmly on the tab.

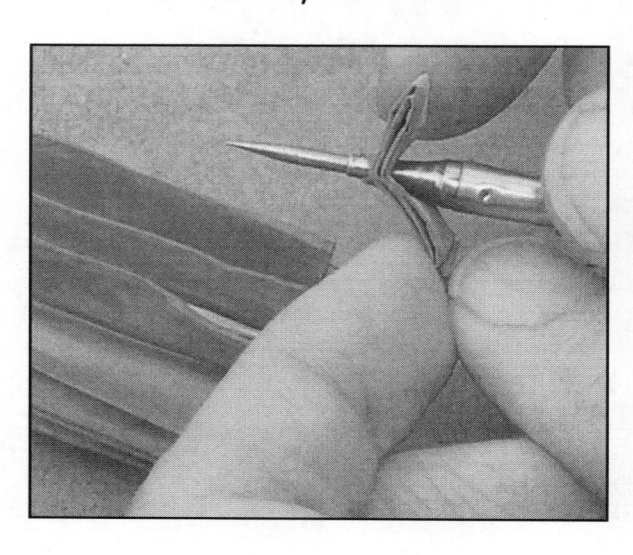

- Make a hole in the folded tab with a drill bit or other small pointed object so that the 8-32 rod will be able to just slip through it.

Fitting the negative electrode assembly

You will need six 8-32 nuts and one stainless steel fender washer.

- Place the rubber washer into the bottom of ¾" PVC pipe cap.
- Place the nickel washer on top of the rubber washer in the pipe cap.
- Screw ¾" pipe cap onto the threaded rod, with washers inside, until about 8" of rod protrudes (measured from inside bottom of cap to tip of rod).
- Slip the negative electrode mesh tab onto threaded rod.

Electrolyzer and Electrolyte

- Slip ⅜" x ⅜" nickel tab onto the rod on top of mesh tab.
- Push the mesh tab and the nickel tab down the rod, keeping the rod centered in the negative electrode.
- Seat the negative electrode in the cap
- Put the stop tab on the threaded rod over the top of the negative electrode.
- Screw on an 8-32 nut until it is firm against the stop tab.
- Observe how much rod is protruding above the nut on the stop tab. Screw the rod until just a little bit of the rod protrudes (about 1/16" to ⅛"). If an extra nut will be used to act as a locking nut for the first nut, add enough space for it.
- Insert the end of the rod that extends through the pipe cap, into the center fitting on the inside of the bottom blind flange.
- From the outside of the blind flange, screw the Teflon insert onto the threaded rod until it bottoms out in the compression fitting. Screw and unscrew the rod to get a tight fit, until you are sure that the Teflon insert is bottoming out in the fitting.

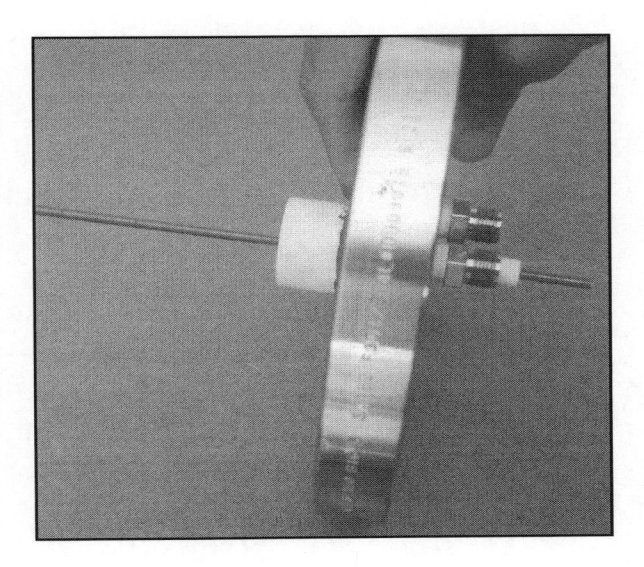

Practical Hydrogen Systems

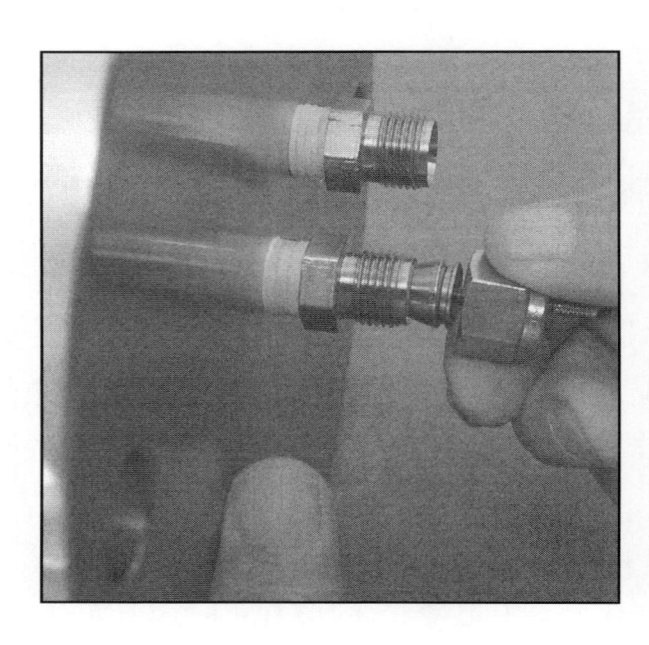

♦ From the outside of the flange, put on the fitting tube nut, and twist it moderately tight so that the ferrules grip the Teflon insert.

♦ Remove the nut, stop tab, mesh electrode and 3/8" x 3/8" tab, but leave the pipe cap and washers, and do not move the threaded rod. Maintain its exact position for marking.

♦ Hold the rod steady and unscrew the PVC cap enough to mark the position on the rod exactly where it comes out of the flange face. This is where the flange meets the bottom of the pipe cap.

♦ On the other side (outside) of the flange, mark the rod at 1" or 1 1/16" from the end of the compression fitting tube. There will be washers and nuts for connectors on this part of the rod.

♦ After marking the rod, unscrew the pipe cap from the threaded rod, being careful not to move the rod.

Installing the negative electrode assembly

♦ Remove the tube nut from the fitting and pull the rod, with the Teflon insert on it, out of the fitting. Do not unscrew the Teflon insert. Maintain the position of the insert on the threaded rod.

♦ Apply thread sealant and several layers of Teflon tape to the threaded rod, from the end of the Teflon insert to the mark that indicates the bottom of the pipe cap.

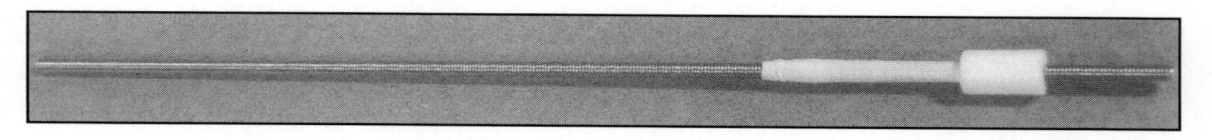

Electrolyzer and Electrolyte

- Back off the Teflon insert by unscrewing it ⅝". Apply thread sealant on exposed rod, then screw the insert back to its original position.
- Cut the rod to size for the external 1" to 1 1/16" for outside connections, per the mark.
- Place epoxy on flange surface that will be covered by the ¾" pipe cap.
- Apply thread sealant for about ⅛" from where tape ends on threaded rod. This will seal the threads on the pipe cap and rod as they are seated.
- Screw the pipe cap down onto flange base and tighten to make sure it is a firm fit.
- Apply epoxy around circumference of pipe cap where it contacts flange.
- Let the epoxy dry for 24 hours.
- Place rod and insert back into the fitting and do a final tightening of the tube nut.

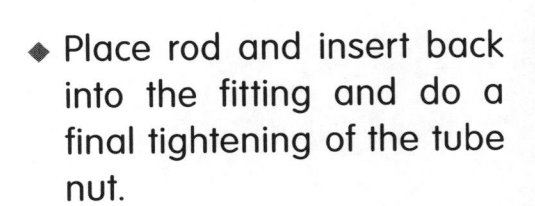

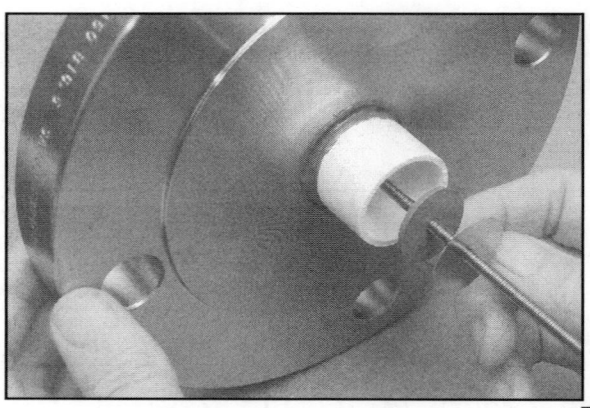

Practical Hydrogen Systems

◆ Slip the rubber washer, followed by the nickel washer, down the rod, and into the pipe cap.

◆ Slip the mesh tab, 3/8" x 3/8" nickel tab, and an 8-32 nut onto the threaded rod. Work these down the threaded rod to the bottom of the cap, with the rod centered in the pleated mesh electrode. Screw the nut tightly to hold the mesh and 3/8" x 3/8" tab firmly down onto the nickel washer.

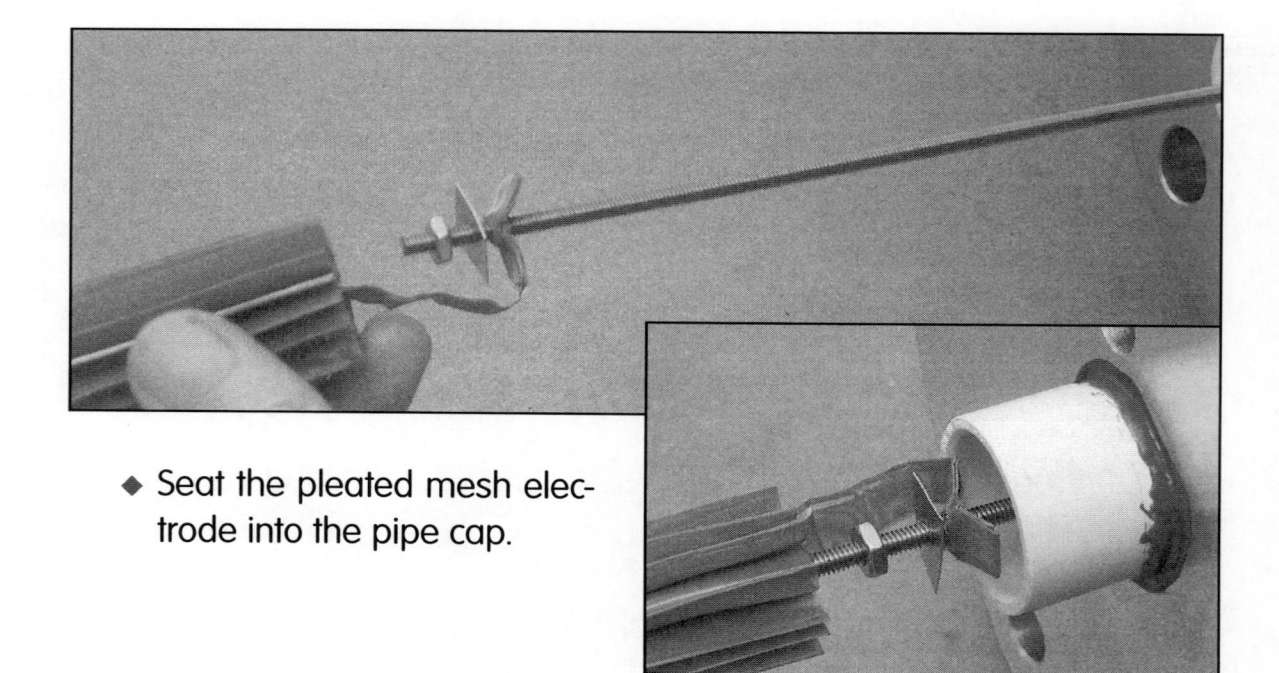

◆ Seat the pleated mesh electrode into the pipe cap.

Electrolyzer and Electrolyte

- Slip on stop tab at the top of the pleated mesh electrode. Bend the two tabs down onto the pleats to hold the mesh electrode firmly in place.

- Screw on retaining nut for the stop tab.

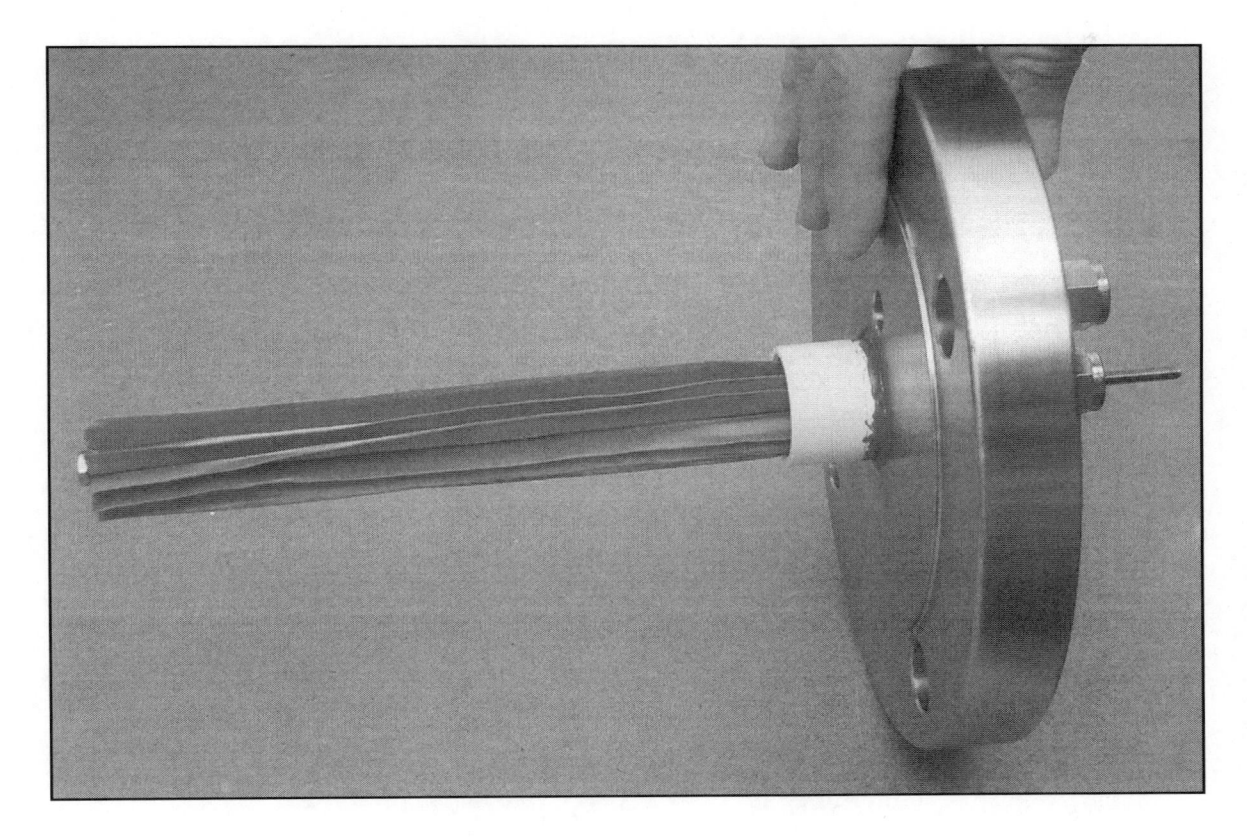

Electrode separator

Fitting the separator

- Cut the polypropylene felt to a 5⅞" x 9⅜" rectangle and roll it into a cylinder 9⅜" long.

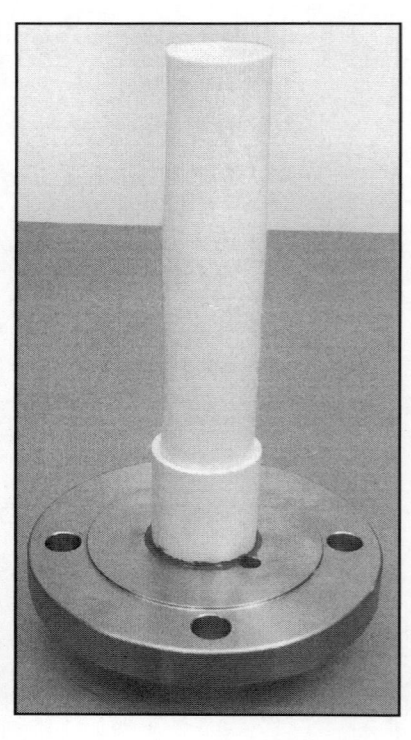

- Fit the felt cylinder into the 1½" PVC coupler. On the inside center of the coupler is a ring of PVC jutting out which the top of the felt cylinder will seat against. The exact length of the felt cylinder will depend on how many turns you were able make with the screw-on flanges. Adjust and trim the cylinder length accordingly. The felt should fit snugly into the coupler and should not be so long that it scrunches up at the bottom of the electrolyzer against the blind flange. There is room for play here. The bottom of the separator should at least be a little below the top of the cap where the negative electrode is seated. It can reach down to the bottom flange, as long as it is not compressed and distorted.

- To test the fit, place the nipple on the bottom flange with a Teflon flange gasket and the negative electrode in place. Place a Teflon gasket on top of the screw flange and slip the felt into the coupler.

Add the bottom flange gasket, then place the bottom blind flange.

Electrolyzer and Electrolyte

- Carefully lower the blind flange/felt separator assembly down over the negative electrode, being careful to keep the electrode centered and to not disturb its position. As you seat the flange and separator, notice whether the felt cylinder is pushing against the bottom flange. If it is, trim the length of the felt to fit.

Assembling the separator

The seam of the polypropylene felt cylinder has to be bonded to complete the separator tube.

Set up a 15 watt soldering iron with a new tip. We used a pointed tip, but other soldering tips, such as a chisel tip, for instance, could be used. A heat gun, or wood burning set element might work quite well also. The best choice would be a heat controlled soldering iron.

Use some felt scraps to practice to find the best temperature for bonding the felt, and to get a sense of how long you can hold the iron on the felt, and how close the strokes should be to each other.

- Roll the felt into a cylinder, as when it was being fitted. With the seam where the two edges of the felt abut in the center, hold the cylinder pressed flat on your worktable. Make sure that the seam edges abut well and evenly.

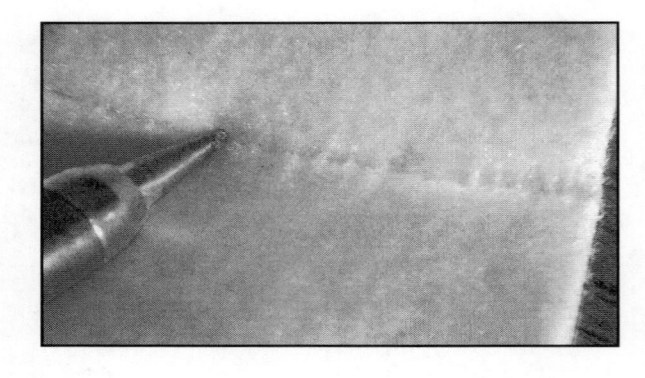

◆ Hold the edges together on one end and apply heat to bond them, by stroking the soldering iron across the seam where the two edges abut. This will melt the plastic and bind the edges together.

◆ Work along the seam, stroking across the edges as you go, and weld the entire length of the tube in this manner. Stroke the iron quite quickly and lightly over the seam. If the iron is held a second too long on the felt, it will melt too much, make a hole, and ruin the piece. Make each stroke with the iron close enough to the previous one so that there are no gaps in the seam.

◆ Hold the piece of felt up to the light to inspect for gaps. If there are gaps, go over the gap area lightly until it is closed. Make sure that no gaps exist because the felt separates the hydrogen and oxygen gases in the electrolyzer, and a leak in the separator would permit the gases to mix, which is dangerous.

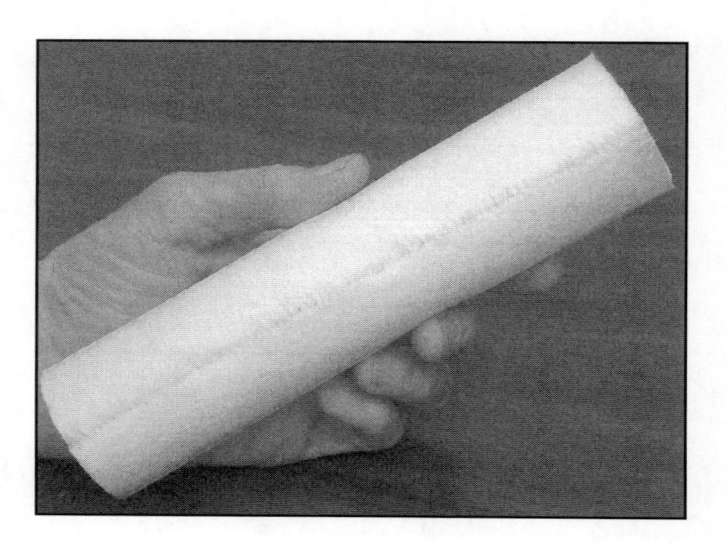

This felt cylinder fits inside the pipe coupler on the hydrogen gas exit port, on the inside of the top flange.

Electrolyzer and Electrolyte

Electrolyzer final assembly

To complete the construction of the electrolyzer, the two blind flanges are attached to the nipple via the screw-on flanges.

Attach the bottom blind flange

- Apply thread sealant to both sides of a Teflon gasket and to the areas covered by the gasket on both the screw-on flange and the blind flange.
- Lay the gasket in place on the screw-on flange
- Insert the blind flange with attached negative electrode assembly into place. Be sure to align the holes in the blind

flange with those in the screw-in flange as you position the flange. The electrolyzer internals need to be right on center – the negative electrode and felt separator will be coming from different directions. Always keep the internal components aligned correctly.

Bolt the flanges

The conventional bolt size for this size flange is ⅝", however the ¾" bolts just fit in the holes and maintain the alignment of the flanges to each other much better. Either ¾" bolts or ⅝" bolts will work.

If using 3/4" bolts

- Drop in two ¾" flange bolts as temporary stays in two holes opposite each other on the flanges, or drop in four ¾" bolts as permanent stays.
- Add nuts to the first ¾" bolts, hand tighten, then add two more ¾" bolts and hand tighten.

If using 5/8" bolts

- Use two ¾" bolts as drop in stays in opposite holes.
- Put two ⅝" bolts in the other two holes.
- Attach the nuts to the ⅝" bolts and tighten to a little more than hand tight so that they will not move.
- Remove the ¾" bolts and replace them with ⅝" bolts and tighten to a little more than hand tight.

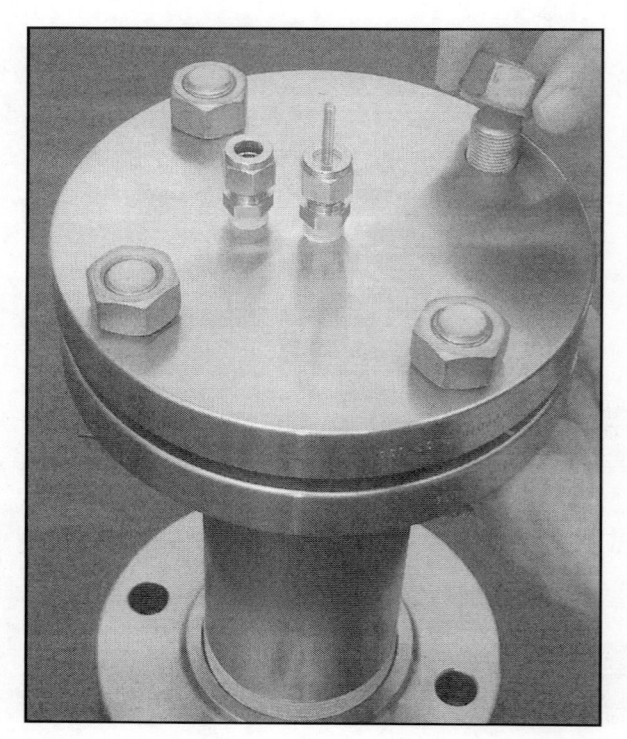

Electrolyzer and Electrolyte

For either size bolts, be sure that the flanges are aligned properly, and continue with the final tightening of nuts and bolts in a crisscross fashion.

- Tighten one bolt.
- Tighten the bolt directly across from it.
- Tighten the other two bolts in the same fashion.
- Repeat this sequence until appropriately tightened.

Do not tighten one bolt completely and then proceed to the next. Just tighten each bolt a little and proceed to the next bolt in sequence. Continue the cycle of bolt tightening until the bolts cannot be tightened any further.

When tightening bolts, use the correct size wrenches. Use two wrenches to tighten the bolts, one to hold the bolt head and the other to tighten.

Attach the top blind flange

The unit thus far assembled has to be turned right side up (with the negative electrode connection on the bottom) to attach the other gasket and the top blind flange. This requires a structure that supports the unit, but does not interfere with or put weight on the external part of the negative electrode on the bottom flange. The electrolyzer unit cannot be put on a flat surface. Some iron or wood supports will suffice for this purpose.

- Prepare the gasket surfaces with thread seal as before, and place the gasket.

Practical Hydrogen Systems

- Place the felt separator into the pipe coupler that is attached to the inside of the blind flange. Make sure it is seated against the center ring on the inside of the pipe coupler.

- Lift up the flange and carefully lower it so that the separator goes down over but does not disturb the negative electrode. The negative electrode ends can catch on the felt if the separator and flange are not carefully aligned and steady as they are lowered into place. As they are lowered, align the holes in the blind flange with those in the screw flange.

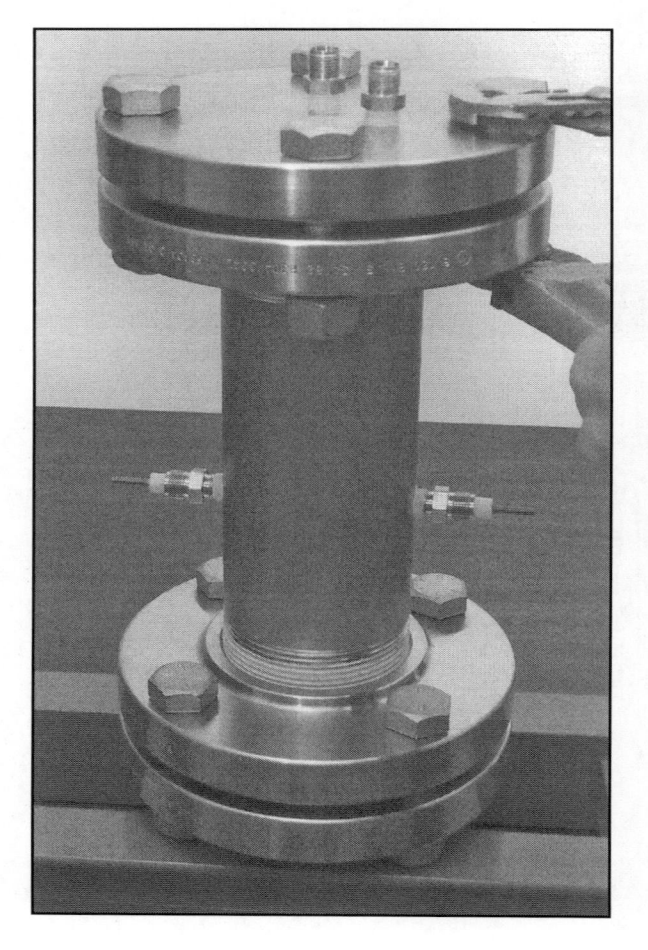

- Fasten the bolts as for the first flange.

- Put all the nuts and ferrules on the four flange ports, so that you know where they are when they are needed for later connections.

The electrolyzer is complete.

Electrolyzer and Electrolyte

Tools for building the electrolyzer

Tool	Supplier	Part #
General purpose HSS screw machine I drill wire gauge size 29, 1 15/16" overall, 15/16" length flute	McMaster-Carr	8947A182
High speed steel hand tap bright finish, bottoming, 8-32, 3 flute	McMaster-Carr	26955A73
TICN coated HSS single end two flute end mill center cut, 7/16" mill dia., 3/8" shank dia, 2" length of cut	McMaster-Carr	30225A42
Heavy duty, light color, cutting and threading oil	McMaster-Carr	2307K26
Oxide treated taper pipe tap for SS high speed steel, 1/4"- 18 NPTF	McMaster-Carr	2661A41
Value-Rite tap wrench, straight handle, style for 0-1/2" (1.6-12.5mm) taps	McMaster-Carr	25605A75
Value-Rite tap wrench, straight handle, style for 1/4"-1" (6-25mm) taps	McMaster-Carr	25605A79
Milling machine or drill press		

Practical Hydrogen Systems

Materials for the electrolyzer

Description	Supplier	Part #	Quantity
1½" PVC pipe coupler, schedule 40, 2⁵⁄₁₆" long, ⅛" thick walls, 1⅛" from internal center ring to edge of pipe opening on either side, 2³⁄₁₆" OD.	Hardware store		
¾" PVC pipe cap, schedule 40, 1" long, 1⁵⁄₁₆" OD, 1³⁄₁₆" ID, ⅛" wall thickness.			
Nickel alloy foil tabs .010 thick, 5 pieces each ½" to ¾" long; and nickel alloy foil washers 1" diameter	McMaster-Carr	#8912K241-4" x 1'	1
Monel standard grade woven wire cloth 200x200 mesh, .0021" wire dia., 12"x12" sheet or 12"x 24" sheet	McMaster-Carr	9225T264 or 9225T34	1
Clear silicon rubber caulking	Hardware store		
304/304L SS welded pipe nipple, Schedule 40, 3" pipe, 10" long, 1⅝" thread length, wall thickness 0.216	McMaster-Carr	4830K361	1
304/304L SS forged blind flange, 150 psi, 3" pipe size x 7½" OD,	McMaster-Carr	44685K118	2
Type 304/304L SS forged flange, 150 psi 3" pipe size x 7½" OD, threaded	McMaster-Carr	44685K18	2

Electrolyzer and Electrolyte

Description	Supplier	Part #	Quantity
Flange gasket kit with bolts and nuts for 3" pipe, 3½" ID, 5⅜" OD gasket, 1/16" thick. Aramid/Buna-N	McMaster-Carr	9166K48	2
(Optional) Virgin PTFE flange gasket ⅛" thick, for 3" pipe, 3½" ID, 5⅜" OD (in consideration)	McMaster-Carr	9483K68	2
(Optional) ASTM A193 grade B8 18-8SS heavy hex head bolt ⅝"-11 thread, 3" length (in consideration)	McMaster Carr	92655A456	8
(Optional) 18-8 stainless steel flange nut ⅝"-11 screw size, 1 1/16" W, 1 3/16" overall H (in consideration)	McMaster-Carr	94758A035	8
(Optional) Polypropylene rod ⅜" diameter, opaque white. Sold in lengths of 8'	McMaster Carr	8658K52	1
(Optional) PVC (polyvinyl chloride) type I rod, ⅜" diameter (in consideration). Sold in lengths of 5'.	McMaster-Carr	8745K42	1
316 stainless steel threaded rod 8-32 thread, 3' length	McMaster Carr	93250A110	1
Teflon rod, ⅜" dia.	McMaster-Carr	8546K12	1 foot
Type 316 stainless steel machine screw nut 8-32 screw size, 11/32" width, ⅛" height	McMaster-Carr	90257A009	sold in packs of 100 only

Practical Hydrogen Systems

Description	Supplier	Part #	Quantity
316 SS Yor-Lok compression tube fitting, male straight adapter for 3/8" tube OD, 1/4" NPT	McMaster-Carr	5182K119	6
Silicone rubber sheet, 3/32" thick, Shore A - 40 durometer, 12"x12" sheet	McMaster-Carr	8632K43	1
Teflon bondable sheet .015" thick, 12"x12"	McMaster-Carr	8711K72	1
Military grade thread sealant tape, premium	McMaster-Carr		
Polypropylene white felt sheeting, 1/16" thick, 72" wide, sold by the foot	McMaster-Carr	88125K11	1'

Parts numbers and suppliers have been provided for your convenience; however, suppliers may go out of business and parts numbers may change. All parts listed are available from multiple suppliers.

Bubbler Scrubber System

Purifying the hydrogen

The purpose of the bubbler in this hydrogen system is to remove contaminants such as particulates and KOH that could be carried to the storage system. These contaminants are scrubbed out of the gas by bubbling the hydrogen through water and/or vinegar. The water absorbs particulates and KOH, and the vinegar helps to neutralize KOH.

Bubbler configurations

The bubbler configuration depends on how much scrubbing is needed for a given system to remove contaminants. Bubbler designs vary and I have noted at least ten different designs used by chemists.

Some configuration possibilities are:

- One bubbler with distilled water
- Two bubblers in sequence with distilled water
- One bubbler with distilled water, and one with vinegar.
- One bubbler with distilled water, and one bubbler with vinegar followed by one bubbler with distilled water.

Bubble breakers

An important goal in bubbler design is to break the larger bubbles into smaller bubbles, which allows more gas surface to be scrubbed. This is usually accomplished with screen mesh of various sizes, or porous media such as aquarium stones. However, if too fine a mesh or porous material is used to break up the bubbles, particulate contaminants will sooner or later clog the fine mesh or the pores of the dispersal stone. If a bubbler is designed for easy disassembly, then the mesh or stone can be cleaned or replaced, and this may not be a concern. Also, fine porous stones and very fine mesh require more pressure to push the gas through them.

This particular bubbler is designed for trouble free service for a long period of time. It has a larger mesh which is less susceptible to clogging. The scrubbing action may not be as effective as it would be with a finer mesh, but it can be permanently enclosed. Also, because of the larger mesh, gross obstructions (though unlikely) can be cleared by simply running water through the bubbler.

There are many porous materials and mesh sizes that are promising and worth trying for improving the scrubbing process.

Other factors that affect the scrubbing process are the distance of bubble travel in the liquid and the volume of liquid in the bubbler. The further the bubbles travel, the more they are scrubbed.

Bubbler liquids

Use only distilled water in the bubbler. If vinegar is used, it should not have sulfur additives. Natural food stores carry vinegar without sulfur additives. Note that some food stores sell vinegar that has "natural" on the front of the label, but if you read the back of the label it may state in small print that sulfur compounds have been added. The main reason to avoid sulfur is that if hydrogen is going to be used for fuel cells and is contaminated by sulfur it will poison the catalyst in the fuel cell; and if a catalytic recombiner is used, the sulfur will quite quickly poison the catalyst and render it inoperable.

Either white or red vinegar can be used. Red vinegar turns lighter in color as it soaks up KOH, which can be a useful visual indicator of when the liquid should be changed.

Note the degree of yellow discoloration caused by the KOH concentrating in the distilled water, or take pH readings to determine when to replace the liquid with fresh distilled water or vinegar. Since the liquid is not visible through the stainless steel walls, unscrew the plug on top of the bubbler and insert a straw a little below the liquid level surface, then cap the end of the straw with your finger. Draw the straw out and release the liquid into a glass and note the color of the liquid, or test the pH with a pH meter, or test paper. There are two outlets, one on the on the top of the bubbler for refilling, and the other on the bottom for draining.

The frequency of draining and refilling depends on the operating hours of the system. If the system is powered by an intermittent source such as solar or wind, this will vary with the weather and season.

We did not include an automatic bubbler refill in our system. This could be easily accomplished by inserting pH probes set to a predetermined value, which would then operate a pump to evacuate and refill the bubbler(s).

When more than one bubbler is used there should be a check valve with a cracking pressure of around 1 psi in between bubblers. This will prevent back-flow from one bubbler to the previous bubbler.

Bubblers can be filled from one half to three quarters of their volume, but they should not be filled to the top. The bubbling action creates a surface spray when the bubbles break at the top of the liquid. This can put aerosols into the gas stream, and if the top of the liquid is too close to the top of the bubbler, the aerosols will cross over into the coalescer, and will dump more fluid in the coalescer than need be. If the fluid level in the bubbler is lower, more of the fluid will be retained in the bubbler.

Additional internal filters

Bubblers can include an internal filter which acts as a coalescer/filter. This can be accomplished by inserting a disk of polypropylene felt or other filter type material across the top of the bubbler on the inside. The filter will trap the fluids and let them drip back down into the bubbler. However, if your bubblers are welded, the welding process will melt the polypropylene, so this kind of internal coalescer is not an option. A substance such as a high temperature, very porous ceramic could withstand the heat of the welding process might be worth a try. Also, if the bubbler was constructed with a screw-on or flange top, the polypropylene could be used. The big advantage of having a coalescer or filter directly in the bubbler is that you could

Practical Hydrogen Systems

then eliminate the external stainless steel coalescer, which is an expensive component.

Constructing the bubbler

The bubbler consists of

- One 12" long, 3"OD stainless steel tube
- One 10" long, ½"OD stainless steel tube
- Two 3" dia., ½" thick stainless steel disks
- Two Yor-Lok ¼" NPT x ⅜" tube, 90° elbow fittings
- Two Yor-Lok compression tube fitting plugs for ⅜" tube
- Two Yor-Lok straight ¼" NPT x ⅜" tube fitting
- Stainless steel woven wire cloth.

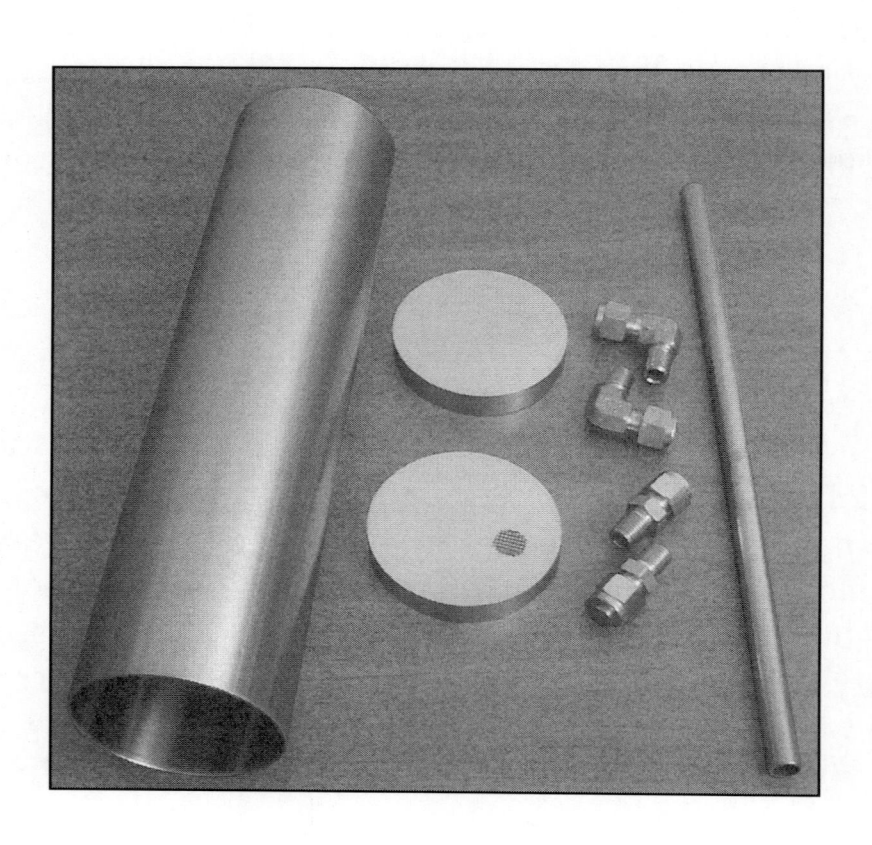

Bubbler Scrubber System

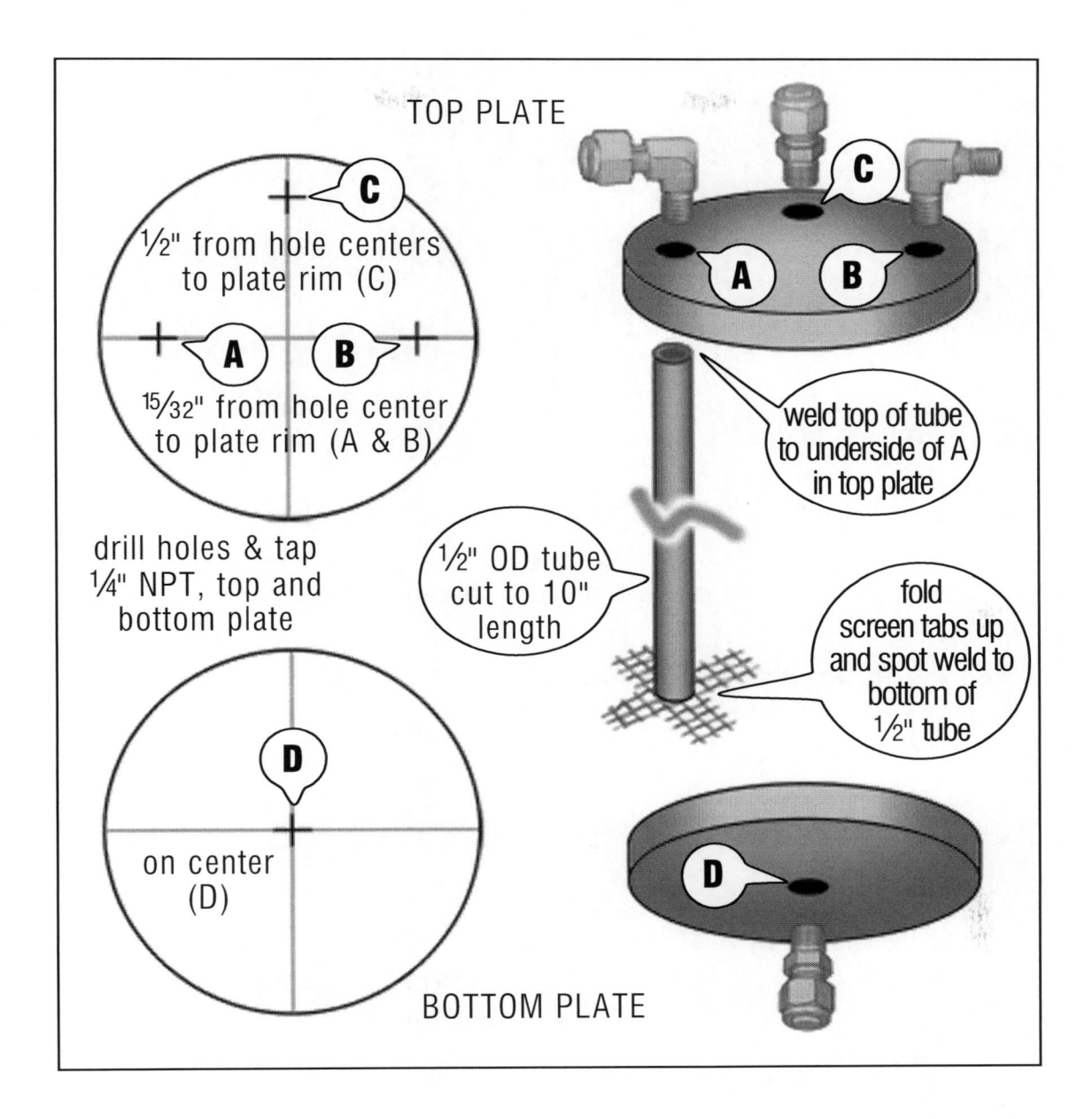

TOP PLATE

½" from hole centers to plate rim (C)

¹⁵⁄₃₂" from hole center to plate rim (A & B)

drill holes & tap ¼" NPT, top and bottom plate

on center (D)

½" OD tube cut to 10" length

weld top of tube to underside of A in top plate

fold screen tabs up and spot weld to bottom of ½" tube

BOTTOM PLATE

Assemble the bubbler

- Drill and tap 3 holes for ¼" NPT fittings in the top disk and one in the bottom disk (see schematics above).

- Cut a piece of ½" stainless steel tubing to 10" length.

149

Practical Hydrogen Systems

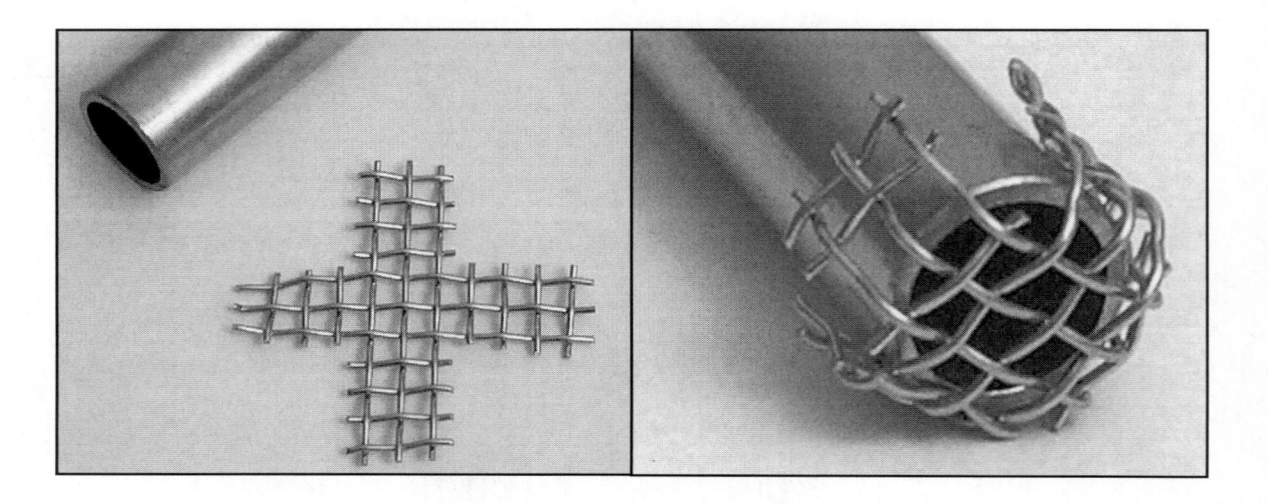

- Cut out a piece of wire screen to fit over end of the tube.

- Scrub and clean all parts in solvent, rinse with water before welding.

- Weld the ½" tube to the top disk.

- Weld the wire screen to the ½" tube.

- Weld the top and bottom disk to the 3" OD stainless steel tube.

- Thread seal Teflon tape, screw in and tighten the two elbows to the top disk.

- Thread seal Teflon tape and screw in and tighten the two straight fittings.

Bubbler Scrubber System

◆ Remove the tubing ends and ferrules from the two straight connectors and replace with plugs. The top plug will act as a fill port and the bottom plug will act as a drain port.

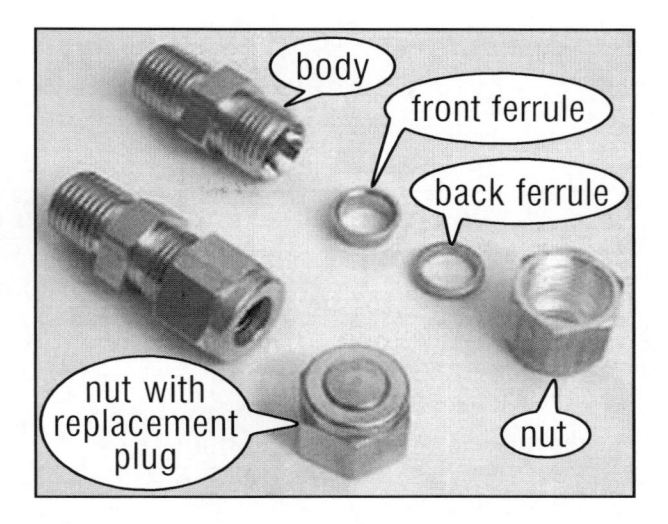

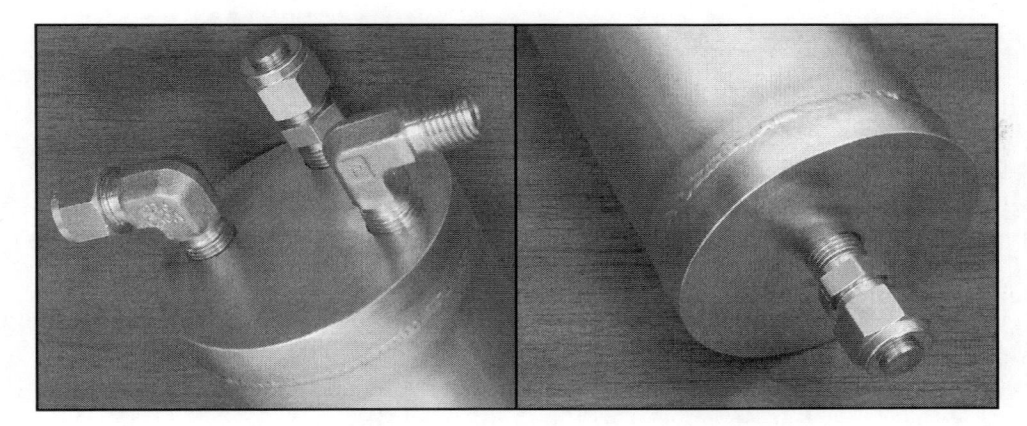

Left, bubbler top. Right, Bubbler bottom

151

Practical Hydrogen Systems

Materials for the bubbler

Description	Supplier	Part #	Quantity
Type 316 stainless steel seamless tubing ½" OD, .402" ID, .049" wall, 1' length	McMaster-Carr	89785K247	1
High-polish stainless steel tubing type 316/316L, 3" OD, 2.87" ID, .065" wall, 1' length	McMaster-Carr	4466K491	1
316 SS Yor-Lok compression tube fitting plug for ⅜" tube OD	McMaster-Carr	5182K626	2
Type 316 stainless steel disc 3" diameter, ½" thick	McMaster-Carr	9260K51	2
316 SS Yor-Lok compression tube fitting male straight adapter for ⅜" tube OD, ¼" NPT	McMaster-Carr	5182K119	3
Type 316 stainless steel woven wire cloth 8x8 mesh, .028" wire diameter, 12" x 12" sheet	McMaster-Carr	9319T538	1
Precision threaded type 316 SS pipe fitting ¼" pipe size, 90° male elbow, 7500 psI	McMaster-Carr	48805K41	

Parts numbers and suppliers have been provided for your convenience; however, suppliers may go out of business and parts numbers may change. All parts listed are available from multiple suppliers.

Connections and Pressure Balancing

Hydrogen side fittings and tubing

This assembly is one of several connecting assemblies in the system. It connects the pressure tank, electrolyzer and bubbler and includes a check valve, a pressure gauge, and an on/off valve. Stainless steel Yor-Lok tube fittings are used throughout with ⅜" stainless steel tubing.

Dimensions are given here for the tube length we used, but tube length will vary for any particular system according to the positioning of each device; and the support stand, enclosure, and base used.

When setting up a system, first figure the minimal distance between components, so that they do not interfere with each other. Determining the optimal length for hard tubing so that the ends of the tubes sit exactly and squarely within the fitting seat is a bit tedious, but it is worth taking the pains to get it right.

Use ⅜" wooden dowel rods and cut them in increments to get the exact length needed for each piece of tubing to seat properly. All of the ends of the dowel rods (and later, the steel tubing) must be cut square so that they seat well in the fittings.

Practical Hydrogen Systems

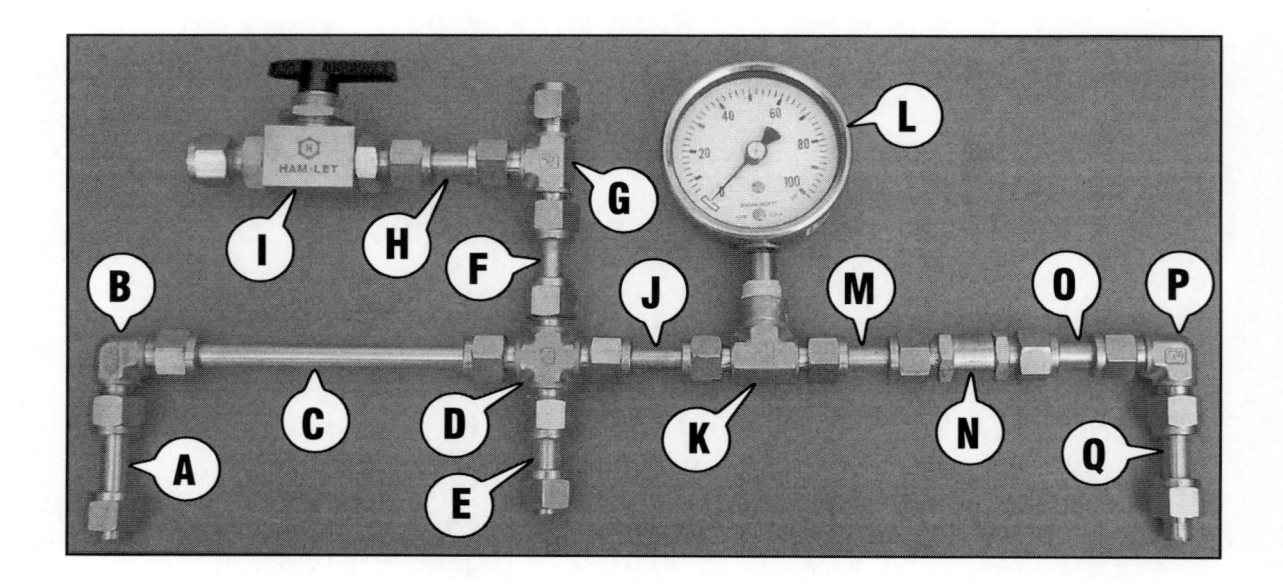

In our system, the sequence of tubing and fittings starting at the pressure tank, then to the electrolyzer and bubbler on the hydrogen side are as follows.

A. Tubing 1⅝" long

B. Elbow ⅜" x ⅜" tube

C. Tubing 4¾" long

D. Cross for ⅜" tubing

E. Bottom of the cross is connected to the hydrogen outlet at the top of the electrolyzer with tubing 1⅝" long

F. The top of the cross is connected to tubing 1⅝" long

G. T fitting, ⅜" x ⅜" x ⅜" tube – the top of the T can be configured for use with either an accumulator, or an equalizer valve. When using the accumulator, the top outlet of the T will be plugged. The plug can be removed and tubing added to connect to the equalizer when desired. If only the accumulator will be used, then an elbow could be used instead of the T, since the extra outlet would not be needed.

Connections and Pressure Balancing

H. Tubing 1⅝" long

I. Valve

J. The right side of the cross is connected to tubing 1⅝" long

K. NPT female T fitting ⅜" x ⅜" x ¼"

L. 0-100 psi gauge is screwed in to the ¼" female top outlet of the T

M. Tubing 1⅝" long

N. Check valve ⅜" x ⅜" tube

O. Tubing 1⅝" long

P. Elbow ⅜" x ⅜"

Q. Tubing 1⅝" long, which connects to the bubbler

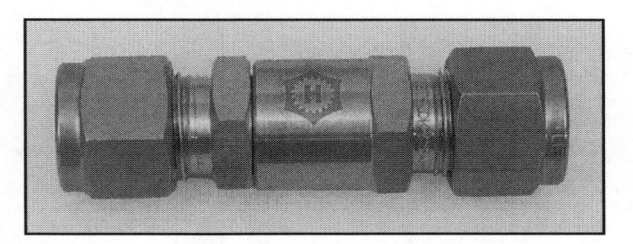

Check valve

Oxygen side fittings and tubing

Tubing and fittings from the top of the electrolyzer on the oxygen side:

- **A.** Tubing 1⅝" long
- **B.** Male NPT ⅜" x ⅜" x ¼" T fitting
- **C.** The ¼" NPT male thread of "B" is connected to the female side of a ¼" female x ¼" female x ¼" male T fitting.
- **D.** Male ¼" side of fitting "C" is connected to the inlet side of an adjustable check valve with two ¼" NPT female ports.
- **E.** Female side of fitting "C" connects to a 0-100 psi gauge.
- **F.** Tubing 4¾" long
- **G.** Male NPT ⅜" x ⅜" x ¼" T fitting
- **H.** Tubing 1⅝" long
- **I.** Valve ⅜" x ⅜" tube

For accumulator J:

- **J.** The male ¼" NPT end of "G" connects to a ¼" female x ⅛" male NPT hex coupler
- **K.** The accumulator "J" ⅛" female connector is screwed on to the ⅛" male side of the hex coupler.

For equalizer valve:

- **L.** Male ¼" NPT end of "G" connects to a ¼" female x ¼" female coupler.
- **M.** Oxygen side of the equalizer valve

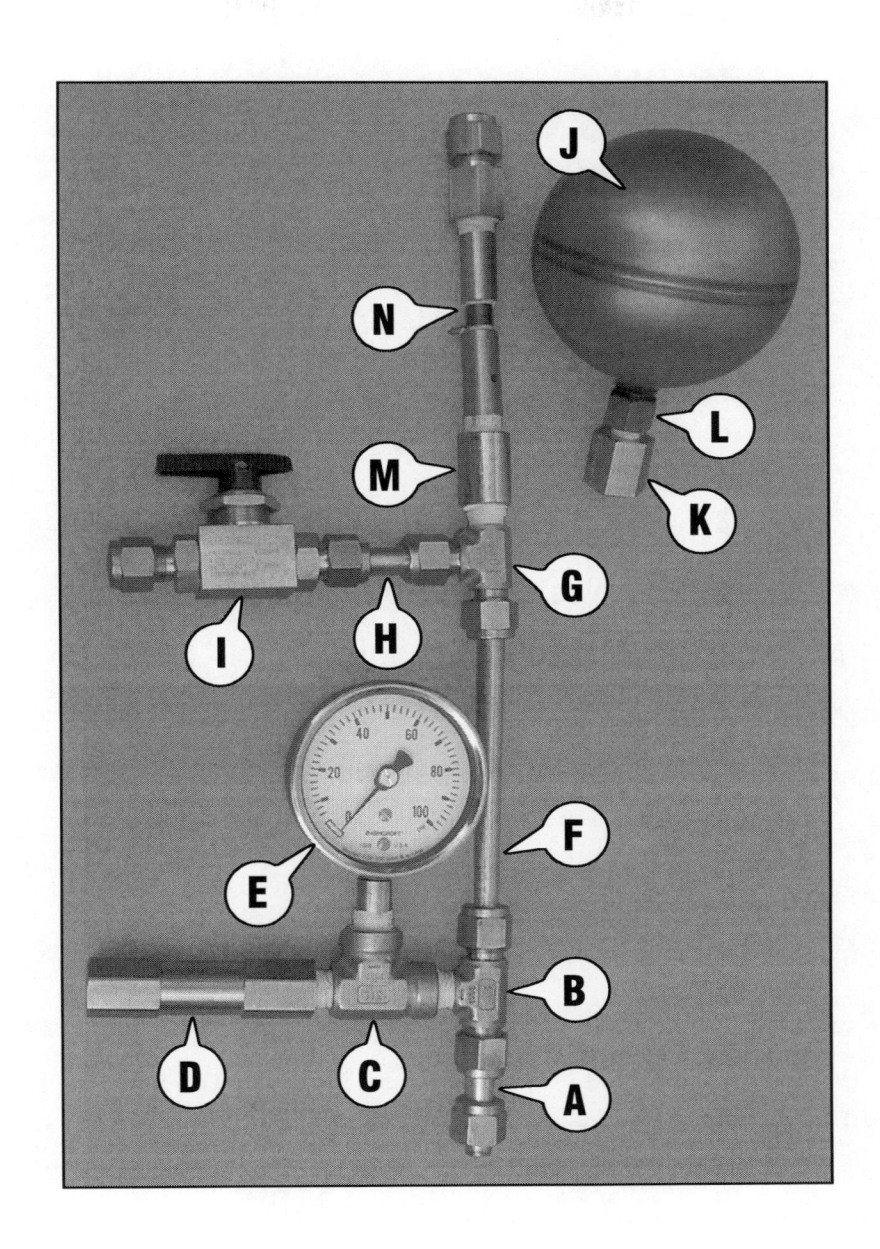

Connections and Pressure Balancing

157

Practical Hydrogen Systems

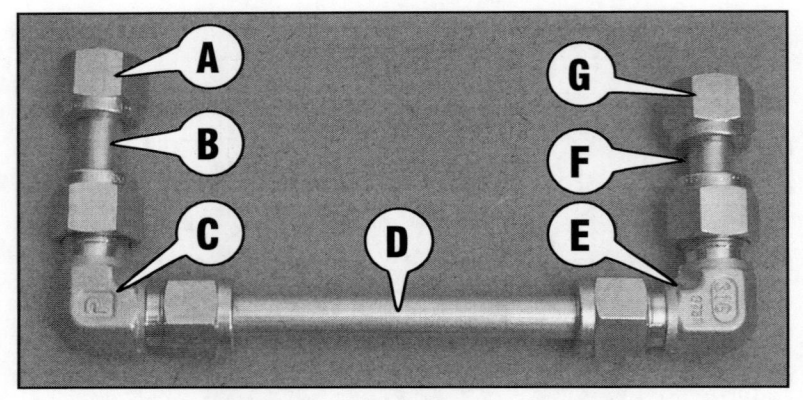

Pressure tank to electrolyzer (left to right) connection, including nuts for electrolyte ports on each end

Electrolyzer to pressure tank connections, bottom

From pressure tank to electrolyzer, left to right:

A. Nut for connection to pressure tank

B. Tubing 1¾" long

C. Elbow ⅜" x ⅜"

D. Tubing 3⅞" long

E. Elbow ⅜" x ⅜"

F. Tubing 2¼" long

G. Nut for connection to electrolyzer

Please note that our tube lengths will vary from yours depending on how you mount your electrolyzer, and all your other specific design changes.

Setting the adjustable check valve (oxygen outlet)

Check valve ranges depend on parameters of the system and its purpose. Some of the ranges available are 3-15, 15-65, and 65-175 psig. We used a 15-65 psig for most of our testing.

Connections and Pressure Balancing

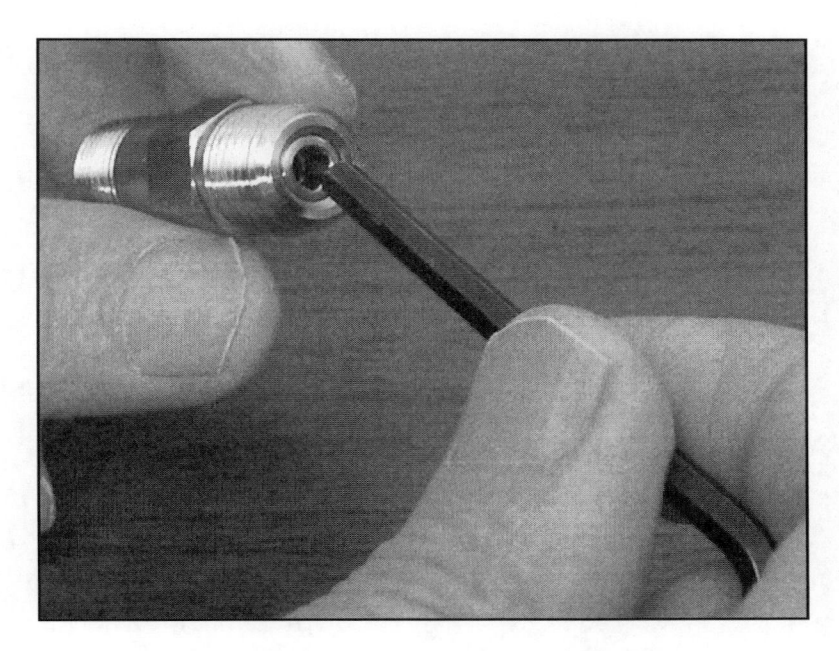

On the outlet side, inside the valve, are two nuts that can be adjusted with an allen wrench. The top nut is the locking nut and holds the nut underneath it in place.

To set the range:

- Attach the inlet of the check valve to a small lab compressor with a pressure gauge attached between the compressor and the inlet of the check valve.
- Turn on the compressor
- Unscrew the top locking nut with an allen wrench
- Push the wrench in a little further so that it engages with the pressure adjustment nut
- Adjust the pressure nut until the desired pressure is displayed on the gauge. To increase pressure you will turn the nut clockwise. To decrease pressure you will turn the nut counter clockwise.
- When you have the bottom nut positioned at the desired pressure, use the allen wrench to tighten the lock nut against the pressure adjustment nut.

Practical Hydrogen Systems

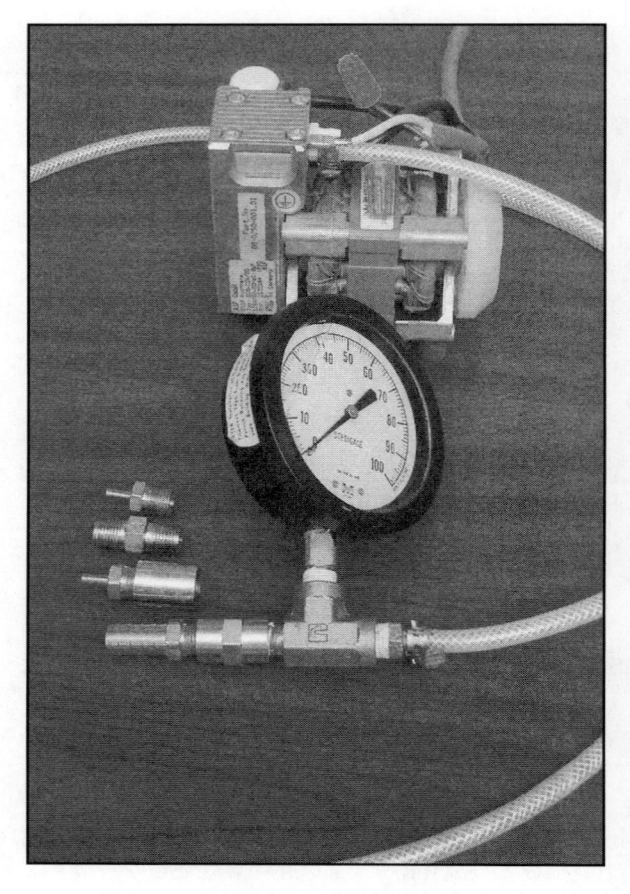

Pressure/vacuum pump with fittings for testing

Repeat as necessary until the desired pressure is achieved. When the locking nut is tightened, it tends to move the pressure adjustment nut, so it may take a few attempts to get it right.

In developing a system, it's a good idea to start out with a low pressure test of around 20 psi. For this operating pressure, set the adjustable check valve to about 25 psi, which is 5 psi above the normal operating pressure. This setting will remain the same whether the accumulator or the equalizer is used. For an operating pressure of 60 psi, the adjustable check valve would be set to 65 psi, and so on.

These settings are not written in stone. As you experiment, you may find other settings more appropriate for your needs. These settings are simply given as a quick reference and starting point.

Over a period of time, some of the liquid in the bubblers is lost into the coalescer and filters, which collect liquid that is regularly drained. Also, some liquid will make it all the way through the system to the storage tank. What is lost through the coalescers and filters, and what is lost to the storage tank, changes the liquid volume in the hydrogen side, and thus, the gas volume. This is minimal over a short period of time, but extended periods of operation require observation of pressure differential via installed gauges to ensure proper operation.

Connections and Pressure Balancing

Gas migration and pressure differential

The electrolyzer produces twice as much hydrogen as it does oxygen. In order to balance the pressure from one side of the electrolyzer to the other it is necessary to introduce several mechanisms to offset this difference. It is important because a gross inequality in pressure will cause the one or the other gases to migrate through the opposite duct in the electrolyzer, and mix the gases. Since we are not storing the oxygen produced by the electrolyzer in this experimental system, a small introduction of hydrogen into the oxygen side is of no concern, save for the safe expulsion of the mixed gases from the system. What needs to be avoided is the introduction of oxygen into the hydrogen side of the system, since the hydrogen will be stored.

The liquid levels of bubblers change over a period of time, which changes the gas volume, so system design should go towards a gas crossover error that favors a minor migration of hydrogen to the oxygen side, rather than oxygen to the hydrogen side.

The accumulator

The oxygen accumulator is a simple pressure balancing device, and is sized according to the total volume of tubing and other components on the hydrogen side. It is an easy and inexpensive way to balance the output of each side of the system within a certain range. The disadvantage is that the volume of liquid in the bubbler affects the pressure balance,

Oxygen side fittings with accumulator sphere

161

so, if an accumulator is used, keep the bubbler liquid levels near the level used to calculate the size of the accumulator.

How accumulators balance pressure

When using an accumulator, the liquid fill level of the bubbler or bubblers will depend on the size of the accumulator. The loss of liquid due to system operation introduces a variable that must be taken into account if an accumulator is used. The bubbler liquid fill level should be maintained with minimal fluctuation in order for the system to operate as intended. As mentioned previously, an intended error should be included into system volume calculations to favor the crossover of hydrogen into oxygen side rather than the other way around. This is achieved by filling the bubbler or bubblers with a little more liquid than your calculations indicate would give a perfect balance. This slightly reduces the gas volume on the hydrogen side, and increases the pressure of the hydrogen.

We operated with a pressure difference of 1 to 2½ psi with the hydrogen side at the higher pressure, and this worked well. The ideal would be to maintain a pressure differential of about 1 psi. At a difference of about 2½ psi, there should be no crossover in the electrolyzer, however, be sure the pressure is always higher on the hydrogen side than on the oxygen side.

Calculating the hydrogen side gas volume

In order to determine the size of the accumulator needed, calculate the volume of space within all the tubing and fittings, coalescers, filters, electrolyzer, pressure tank and sight glass that will contain hydrogen gas during operation. Everything should be included except the storage tank. It is not necessary to get a dead-on accurate figure for this – you just need a ballpark figure.

Connections and Pressure Balancing

Calculating the oxygen side gas volume

On the oxygen side, the volume of the oxygen tubing, and the portions of oxygen duct inside the electrolyzer that will contain oxygen (above the liquid level line), when added to the volume of the accumulator should be half the volume of the hydrogen side of the system.

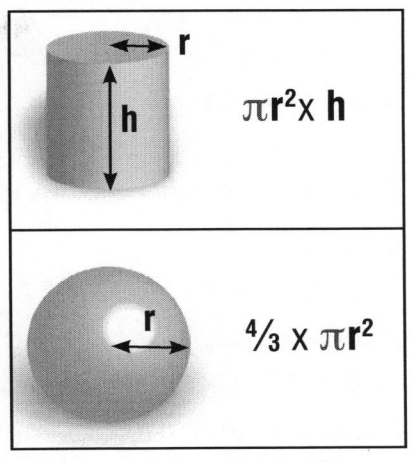

$\pi r^2 \times h$

$\frac{4}{3} \times \pi r^2$

Formulas to calculate volume

With a general idea of the volume needed for the accumulator, you can then look for off-the-shelf parts, or fabricate something. If you find an off the-shelf part that comes close, the volume on the hydrogen side can be adjusted somewhat by varying the liquid level of the bubbler system, as discussed previously. However, do not fill bubblers more than ¾ full with scrubbing liquid.

Fine tuning liquid levels for the accumulator

For this particular accumulator sphere, we used 2.5 cups of distilled water in the bubbler. The amount of liquid will, of course, vary for different accumulators, bubblers, systems, and for smaller pressure differentials, for instance.

Test the system with a bubbler liquid level based on your calculations, then note the pressure differential and adjust the bubbler liquid level accordingly. This may sound tedious, but it is really very simple and easy to do – it works well, and permits the use of inexpensive off-the-shelf parts.

A better method is to use an equalization valve, but we wanted to explore simpler alternatives. One option is to use a differential pressure switch with solenoid valves that will open or close to balance the system. The valves are readily available, but this is actually a more expensive option because hazardous environment and chemical and material compatibility requirements must be taken into account.

Accumulators can be purchased or you can fabricate your own. We used a 3.5" diameter high pressure stainless steel sphere. This was a larger size than was ideal for the system, and it required that the bubbler have less liquid than we would have preferred – the more liquid in the bubbler, the better it is for scrubbing. However, it was a nice looking sphere lying around begging to be used in a project, so we used it.

Other types of accumulators

We also constructed other accumulators from pipe nipple and stainless steel reducing couplings that also worked quite well as you could interchange different lengths of pipe nipple to change the volume easily. Basically any container with the appropriate fittings, and rated for the pressures you will be working with will do.

Another type of accumulator could be made from coiled tubing cut to the appropriate length that would add the volume needed for the oxygen side. If you start with a test pressure of about 20 psi, you can use plastic tubing to act as the accumulator. This

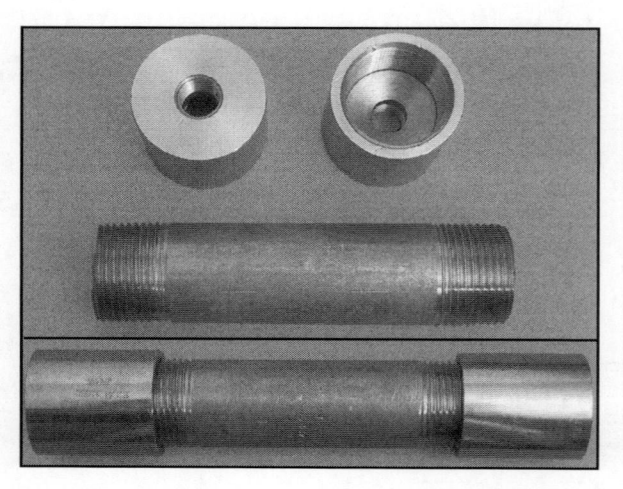

Accumulator constructed from pipe nipple and couplings

accumulator can be tuned to the pressure needed by applying a pinch valve to different locations on the coil. Test the system in running mode, and the pressure gauges will show when you have the exact pressure balance needed. Pre-run volume calculations will determine roughly the length of tubing needed, then the tubing can be fine-tuned with a pinch valve while watching the pressure gauge readings.

Once you have a general idea how much variation occurs over a period of time, replace the coil of plastic tubing with a coil of stainless steel tubing with hard line valves that will give different volumes as needed.

Connections and Pressure Balancing

Using the plastic coil is a good quick technique to calculate the exact volume needed for the accumulator on the oxygen side to balance out the hydrogen side. Although one can calculate "on paper" the volume needed there's bound to be some error. Starting with the plastic coil technique will give more exact results. The tube volume values can be quickly changed by moving the pinch valve and observing the effect. Once you have the desired pressure reading, calculate the volume of the tubing exactly, then construct an accumulator of to match those volume dimensions.

The plastic tube technique can also be used to mitigate the variable shift that occurs with liquid loss in the bubbler system when it is not refilled. In this case, simply move the pinch valve to the appropriate spot to rebalance the system.

For testing at very low pressures the pinch valve and plastic tubing works, but at higher pressures tubing must be stiffer and reinforced; and a pinch valve may not be able to maintain pressure without leaking.

Another interesting possibility is using a hybrid accumulator tank and coil system.

Equalization valves

Another approach we used to balance the pressure differential of the oxygen side and the hydrogen side output of the electrolyzer was an equalization valve. The equalization valve is a simple device that uses opposing pressures to determine which gas is released and how much is released.

In our system, without an oxygen accumulator, the volume of the oxygen side is less than half of the volume of the hydrogen side. If the oxygen side volume was exactly half of the volume of the hydrogen side, the oxygen and hydrogen pressure within the system would be balanced. Since, without the accumulator, the volume available vfor oxygen is significantly less than the ideal balanced volume, initially the oxygen pressure will build up faster than the hydrogen pressure.

Practical Hydrogen Systems

How the valve works

The equalization valve addresses the pressure disparity with a miniature hydraulic cylinder and a few machined parts. The miniature hydraulic cylinder is the heart of the valve. It responds to both gases to determine the amount of gas released from the oxygen side, based upon the pressure of both gases in relation to each other. The cylinder opens and closes an outlet port for the oxygen.

When the pressure on the oxygen side rises higher than the hydrogen pressure, the oxygen gas pushes in one direction against the cylinder, which is being pushed in the other

Our simple equalization valve

Connections and Pressure Balancing

direction by the hydrogen gas. Oxygen is released until the gases have equal pressure. The pressure on the hydrogen side acts as the pilot and determines how much oxygen is to be released from the system.

Constructing a simple equalization valve

Very few parts are necessary to construct this valve:

- Brass rod, ½" diameter
- Miniature cylinder
- NPT fittings, ¼" to attach the equalizer to the rest of the system
- ¼" diameter silicone rubber disk, 3/32" thick

The valve is easy to fabricate using engraver's brass, which is quite soft. See the illustrations on the next page for machining instructions, and photos at left on page 166 for assembly. This is a prototype valve and not intended for use within a system for any length of time, due to the use of the brass rod. If you wish to skip the valve testing phase, use stainless steel rod rather than brass rod. Our cylinder is made from nickel plated brass with a stainless steel piston, so that part of the valve is fine for long term use. If you do not have a small milling machine, you can have a machine shop fabricate it.

After the holes are drilled and the rod is threaded, simply screw the standoff onto the threaded piston rod after applying a little sili-

Practical Hydrogen Systems

A — 1/2" diameter brass rod, 1 1/2" long
B — 1/8" hole on center all the way through
C — 11/32" hole on center, 1/2" deep
D — tap 3/8"- 32, 1/2" deep
E — 1/4" NPT

HYDROGEN SIDE

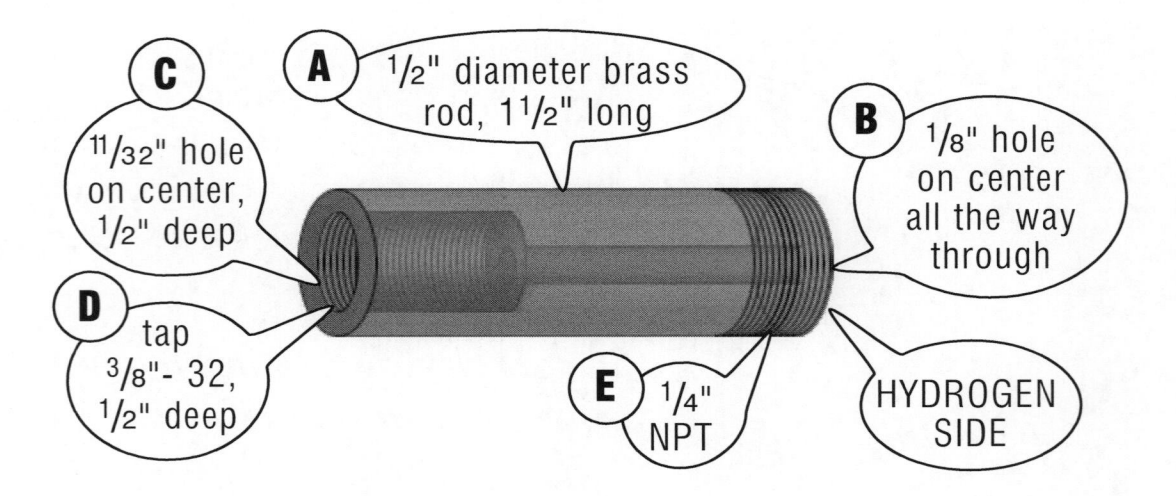

cone or epoxy glue. Punch a 1/4" diameter silicone rubber disk from a 3/32" thick piece and glue this onto the end of the standoff. This is the gasket-seal for the oxygen outlet.

This is only one particular design among many that can be used and no doubt many improvements can be made on this particular design.

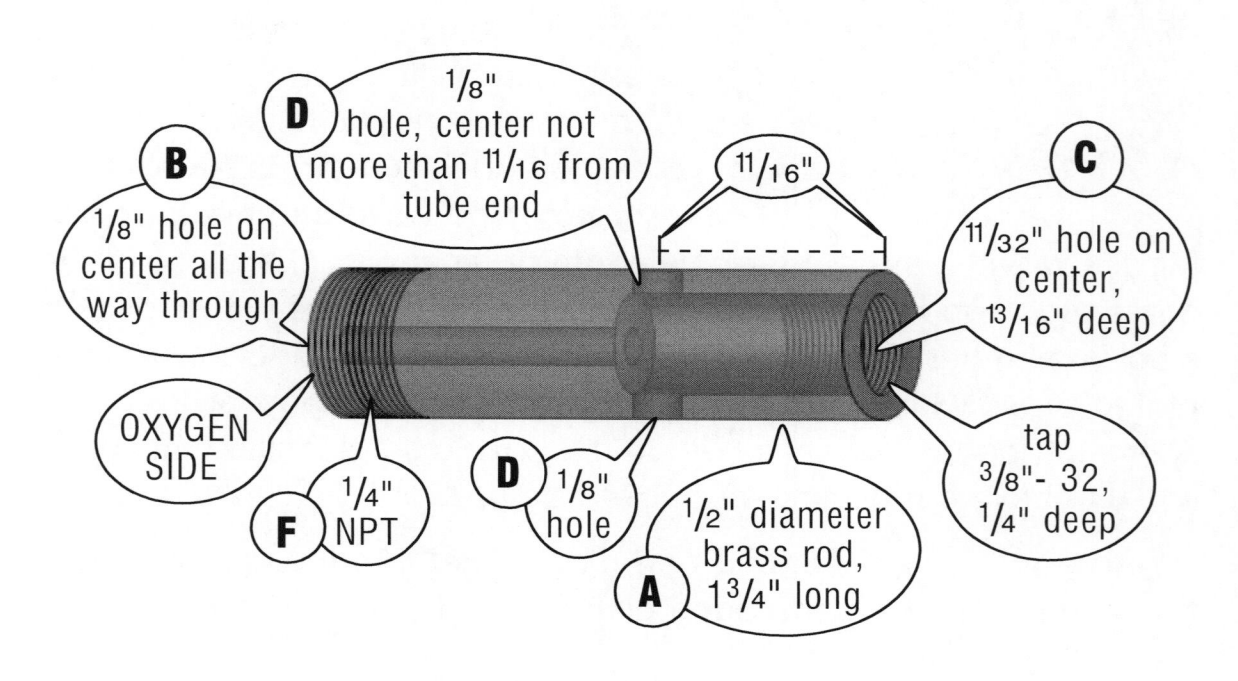

A — 1/2" diameter brass rod, 1 3/4" long
B — 1/8" hole on center all the way through
C — 11/32" hole on center, 13/16" deep
D — 1/8" hole, center not more than 11/16 from tube end
D — 1/8" hole
F — 1/4" NPT
tap 3/8"- 32, 1/4" deep
11/16"

OXYGEN SIDE

168

Connections and Pressure Balancing

Equalization valves with bellows

Our equalization valve did not have a bellows attached, and worked extremely well during the testing period. However, a bellows should be considered a necessary addition to prevent leakage of hydrogen, and to isolate the hydrogen stream from possible contaminants that arise from the cylinder lubricant/seal used.

Stainless steel stock bellows can be purchased from the suppliers listed. Inquire what type of necks are available in stock bellows so that they can be easily attached. Bellows can be attached using epoxy. Most are attached with micro welding or ion beam welding techniques; however, the cost for welding is not justified when epoxy will make a bond that is just as strong and perhaps stronger in some cases.

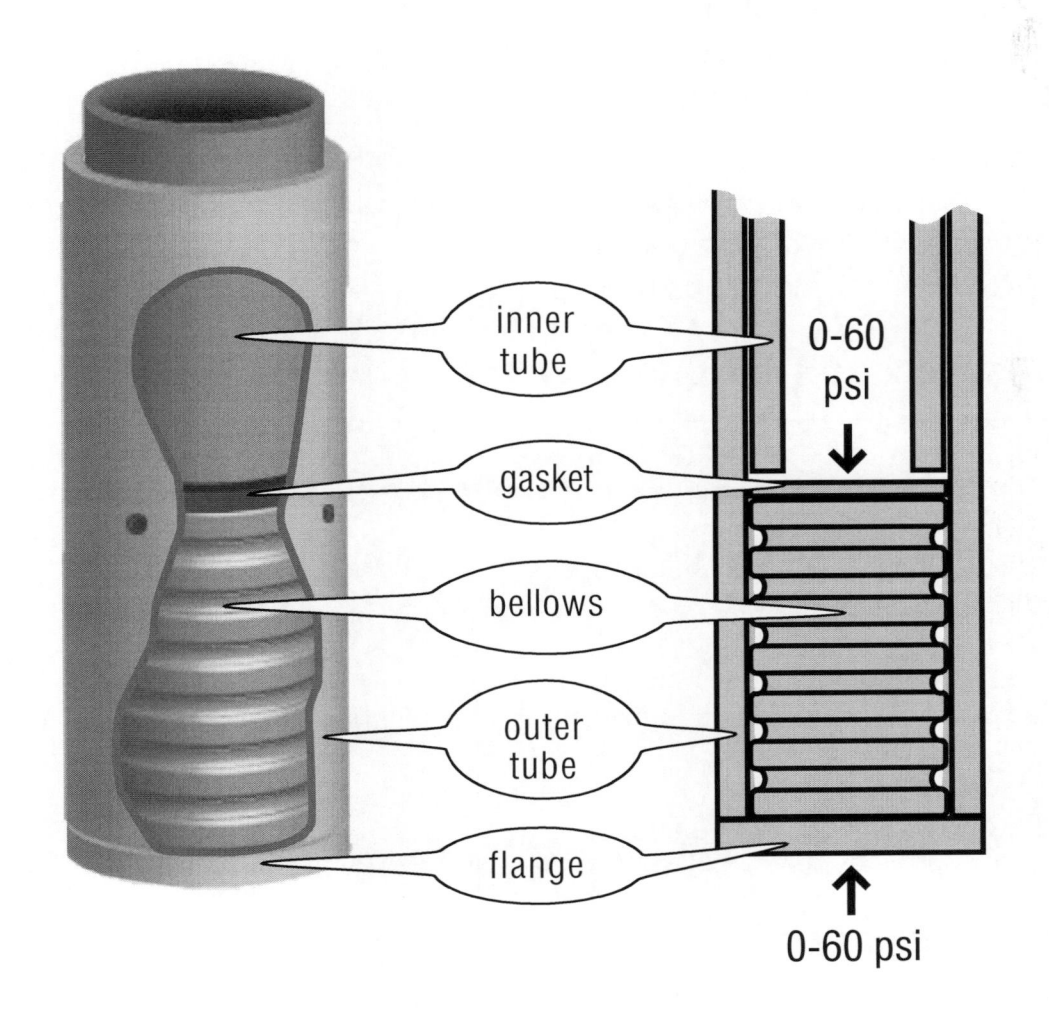

- inner tube
- gasket
- bellows
- outer tube
- flange

0-60 psi ↓

0-60 psi ↑

Other equalization options

- **Electronic differential pressure control.** There are many different systems and components available. Supplier catalogs will provide the information needed to design a system with these devices.
- **Shuttle type valve.** These are usually not that tight and would require experimentation to get something that works.
- **Pilot type valves.** Might work in this application without any alterations, but research the suppliers catalogs.

There are plenty of alternatives available for pressure equalization, but the challenge is to come up with the least costly alternative that will do the job adequately and safely. A particular valve may have some characteristics that you need, but may be missing other necessary characteristics. For instance, it is easy to find electronic/solenoid systems for handling pressure and flow, but then the component must also work with hydrogen and oxygen and potassium hydroxide. The possibilities begin to narrow and the prices rise sharply.

Mechanical valves piloted pneumatically, or electrically driven solenoid valves constructed to operate in hazardous atmospheres will do the job adequately. The issue is one of price, what your budget will allow, and your capacity for ingenious innovation.

Tools for connections & pressure balancing

Tool	Supplier	Part #
High speed steel two flute end mill $^{11}/_{32}$" mill diameter, $^{3}/_{8}$" shank diameter, $1\frac{1}{2}$" length of cut.	McMaster-Carr	3051A59
Fractional size high speed steel hand tap taper, $^{3}/_{8}$" - 32 H3 pitch diameter, 4 flute.	McMaster-Carr	2521A419
Tin coated jobbers twist drill bit, heavy duty, $^{1}/_{8}$" size, $2\frac{3}{4}$" long overall, $1\frac{5}{8}$" long flute.	McMaster-Carr	29115A71

Connections and Pressure Balancing

Materials for connections & pressure balancing

Description	Supplier	Part #	Quantity
Type 304 SS-case gauge 2% midscale accuracy 2½" dial, ¼" NPT male bottom, 0-100 psi	McMaster-Carr	4066K413	2
316 SS spring loaded poppet check valve ⅜" tube OD, Yor-Lok fittings	McMaster-Carr	45385K55	1
Type 316 stainless steel seamless tubing ⅜" OD, .305" ID, .035" wall	McMaster-Carr	89785K837	1- 6 ft. length
316 SS ball valve with Yor-Lok fittings, 2-way panel-mount, straight, for ⅜" tube OD	McMaster-Carr	45775K58	2
Generant adjustable check valve, ¼" female x female (valve range will depend on your operating pressure)	Lindco Inc.	ACV-4FF-SS-B 20	1
Accumulator sphere, stainless steel, 3.5", ⅛" NPT female fitting. Other stock size spheres are available or can be fabricated to specifications.	Aerocon Aerospace		
316 SS Yor-Lok compression tube fitting T, tube x female x tube for ⅜" tube OD, ¼" NPT	McMaster-Carr	5182K323	1
316 SS Yor-Lok compression tube fitting plug for ⅜" tube OD	McMaster Carr	5182K626	1
316 SS Yor-Lok compression tube fitting 90° elbow for ⅜" tube OD	McMaster-Carr	5182K416	4

Practical Hydrogen Systems

Materials for connections & pressure balancing

Description	Supplier	Part #	Quantity
316 SS Yor-Lok compression tube fitting cross for ⅜" tube OD	McMaster-Carr	5182K768	1
316 SS Yor-Lok compression tube fitting T for ⅜" tube OD	McMaster Carr	5182K436	1
316 SS Yor-Lok compression tube fitting T tube x female x tube for ⅜" tube OD, ¼" NPT	McMaster-Carr	5182K323	1
316 SS Yor-Lok compression tube fitting T, tube x male x tube for ⅜" tube OD, ¼" NPT	McMaster-Carr	5182K224	1
316 SS Yor-Lok compression tube fitting T, tube x tube x male for ⅜" tube OD, ¼" NPT	McMaster-Carr	5182K187	1
Precision threaded type 316 SS pipe fitting ⅜" pipe size, female x female x male T, 5000 psi	McMaster-Carr	48805K59	1
Precision threaded type 316 SS pipe fitting ¼" x ¼" pipe size, hex coupling, 6600 psi	McMaster-Carr	48805K69	1
One equalizer			
Extreme-pressure 316 SS threaded pipe fitting ¼" female x ⅛" male, adapter, 6000 max psi	McMaster-Carr	51205K187	1
316 SS Yor-Lok compression tube fitting female straight adapter for ⅜" tube OD, ¼" NPT	McMaster-Carr	5182K266	1

Connections and Pressure Balancing

Description	Supplier	Part #	Quantity
(Optional) Differential pressure switches	Dwyer Instruments Inc. and other manufacturers.		
Miniature cylinder, double acting, ¼" bore, ¼" travel	McMaster Carr	6604K21	1
½" OD brass rod, alloy 353 engraver's brass	McMaster Carr	8313K143	1 3' length
Stainless steel threaded round standoff. ¼" OD, ⅜" length, 6/32 screw size	McMaster-Carr	91125A443	1
Silicone rubber sheet, 3/32" thick, Shore A - 40 durometer	McMaster Carr	8632K43	1 - 12" x 12" sheet

Parts numbers and suppliers have been provided for your convenience; however, suppliers may go out of business and parts numbers may change. All parts listed are available from multiple suppliers.

Coalescers

The coalescer absorbs fine aerosol contaminants, such as water vapor and particulates, and thus removes finer impurities in the gas stream that are not wanted in the final product.

The best material to use for a coalescer in this type of system is stainless steel. The stainless is preferred because of the corrosive effect of KOH. Coalescers are really very simple, with glass fiber, polypropylene or other type of compatible material that filters out the aerosols, which then coalesce and settle at the bottom of the bowl. The bowl-shaped bottom of the coalescer has a drain nut that can be operated manually to drain the water from the bowl; or it can have a spring or lever that automatically drains the bowl when the liquid reaches a certain volume.

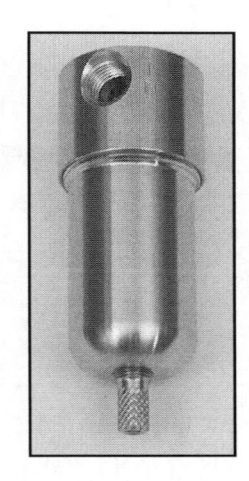

Coalescer

Coalescers come in a variety of shapes and sizes. Larger coalescers don't have to be drained as frequently, which is a factor to consider when a manual drain is involved. With an automatic drain the size is not of major significance, though it is important not to oversize any coalescer in relation to the system involved.

The particular coalescer we used is a small one, but has proven adequate for this system. At first we thought that it might be too small, but the accumulation was a lot less than we expected.

We have placed the coalescer right at the bubbler, but it probably would be better to place a filter at this point, and have the coalescer after the filter.

Coalescers vs. filters

Another option would be to place several filters in line. Generally the term "filter" refers to a component that removes larger particulates and moisture compared to filters that are called "coalescers." "Filters" generally won't catch the fine aerosols that the coalescer will trap.

You can design and build coalescer and filter bodies, and filters. Stock parts can be used with a little drilling and tapping. Filter material can be polypropylene or glass cloth. Do not use organic fiber materials for filters in these systems – they are not compatible with KOH.

Our coalescer is simply screwed on to the 90° Yor-Lok elbow on top of the bubbler after the fitting is thread sealed and taped.

The filter used has to be replaced occasionally. They do last for quite a while, but they are filters and the nature of a filter is to clog up over time as it collects particles. How much a system is operated and how well the KOH and/or distilled water feed is kept free from particle contamination are some of the key factors here. Observe over time how the filter is holding up to get an idea of how often the filters should be changed.

Sintered stainless steel or stainless steel mesh filters are useful in highly contaminated gas streams as they can be cleaned and reused. These filters can be had for 1, 10, 25, and 100 micron efficiencies.

High efficiency coalescing filters are able to remove 0.1 micron particles and aerosols with about 95% efficiency. The particular coalescer for this system has a borosilicate glass element.

Moisture in the hydrogen gas stream

The most common impurity in gas processing systems is water. In addition to filters and coalescers, driers and moisture traps with molecular sieves can be used. For general purpose applications, however, a coalescer and/or filter will suffice.

If the hydrogen gas produced is for fuel cells, it can be beneficial to retain a degree of moisture. The hydrogen side of the membranes of the fuel cells

Coalescers

need to be adequately hydrated. Commercial bottled hydrogen gas can be too dry, and it is expensive to have to add a moisturizing unit.

If the gas production is a supply for recharging hydride bottles, the gas should be as dry as possible. Moisture deteriorates the performance of the hydrides in the bottle. If you use the hydride bottles to supply a fuel cell system, it may be necessary to re-moisturize the hydrogen by running it through a bubbler.

At any rate, it is possible to control the amount of moisture in the hydrogen that you store.

Description	Supplier	Part #	Quantity
Stainless Steel coalescing filter ¼" pipe, 8 scfm max @100 psi	McMaster-Carr	4300K11	1
Precision threaded type 316 SS pipe fitting ¼" pipe size, 90° male elbow, 7500 psi	McMaster-Carr	48805K41	1

Parts numbers and suppliers have been provided for your convenience; however, suppliers may go out of business and parts numbers may change. All parts listed are available from multiple suppliers.

Storage Tank and Peripherals

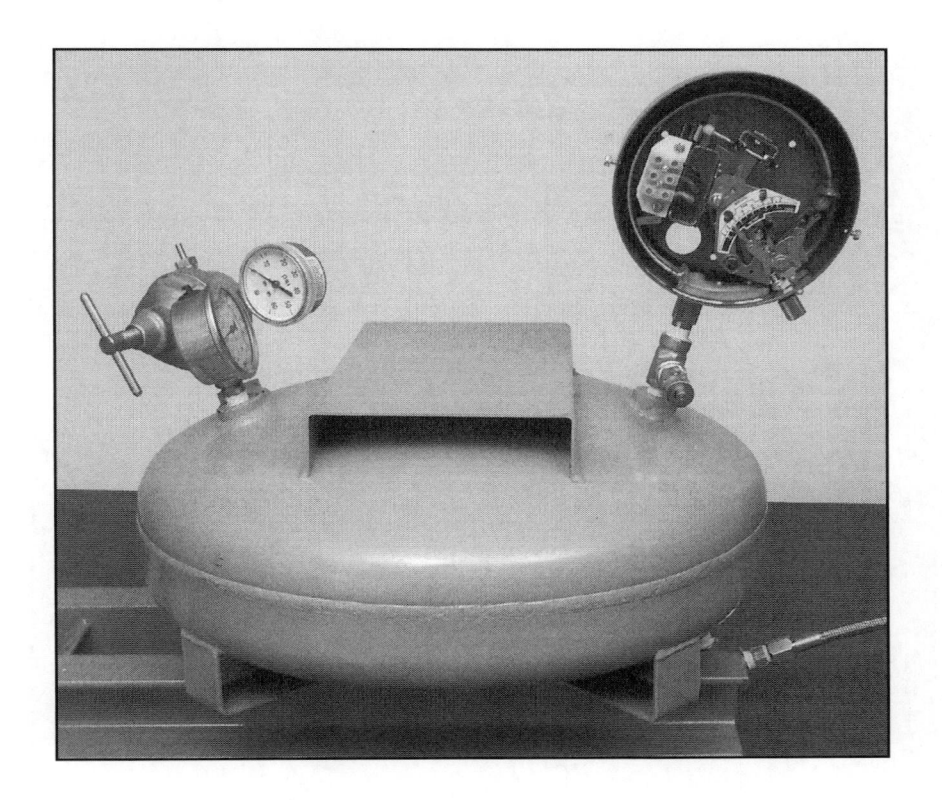

There are many types of tanks that can be used. The best for this type of system is a stainless steel 316 tank. The size of the tank will depend on your storage needs. For initial experimental purposes a small tank is recommended. As discussed earlier, using a smaller tank makes the system's fill cycle shorter and faster, which saves a lot of time when running tests.

Pressure switch

There are three wire connections to be made on the pressure switch. Notice that the switch has three colored coded wires:

- Blue – "High" wire
- Red – "Low" wire
- Black – "Common" connection

Practical Hydrogen Systems

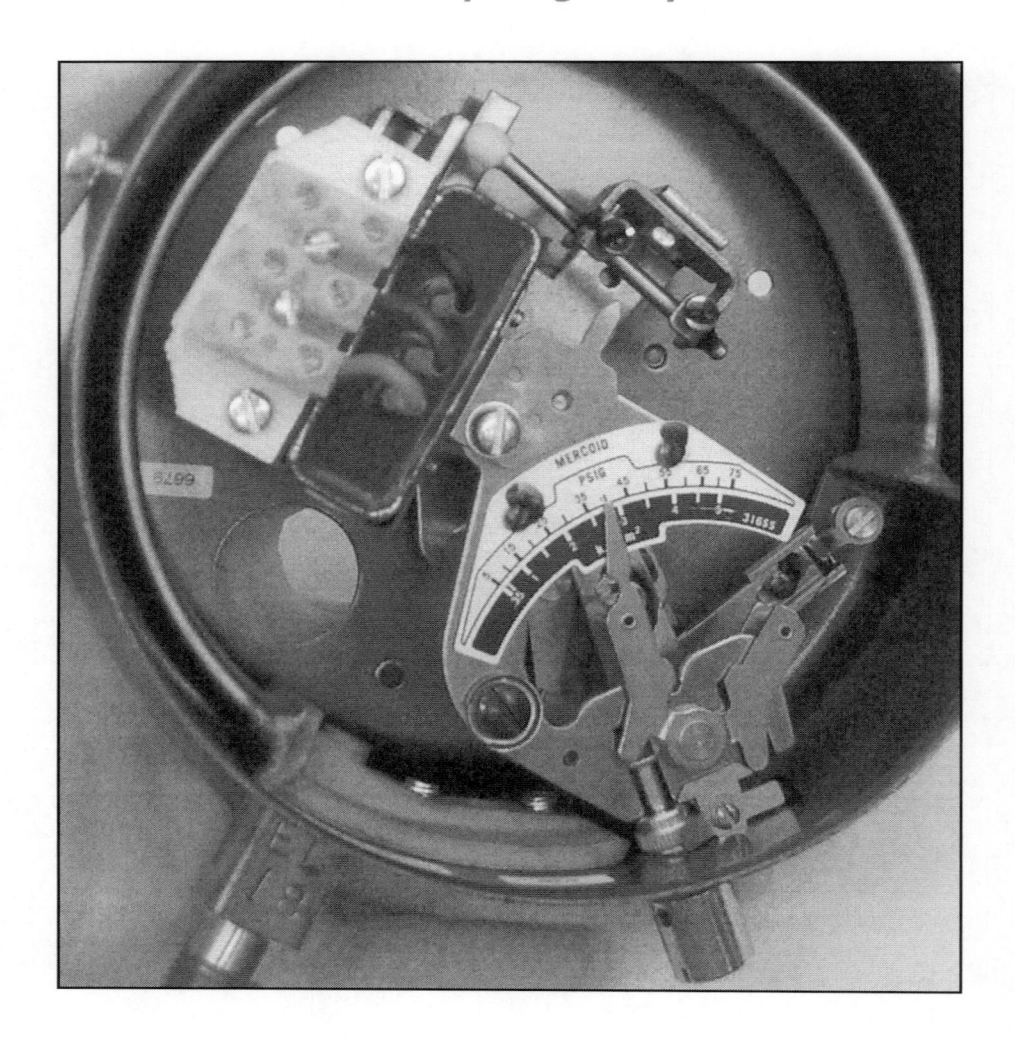

When the tank is filled to the preset maximum pressure, the blue High wire is connected via the switch to the black Common wire. This turns the electrolyzer off. When the pressure falls to a certain preset low pressure the red Low wire is connected to the black Common wire and the electrolyzer is turned on to produce more gas. If there is no power available when the pressure is at the Low preset level, such as when ESPMs or other intermittent energy sources are used, the circuitry will set and be ready to go when power is available – when the sun rises in the morning, or the wind blows again.

Pressure switch characteristics

This particular pressure switch is a mechanical switch that works by connecting a bourdon tube to a lever switch. The bourdon tube is a slender container which moves according to the pressure of the gas entering the tube. The lever switch is preset by mechanical positioning to respond when the pressure rises to a certain point or falls to a certain point. As the pressure rises to the preset high point and falls to the low point, the moving tube pushes against the lever or recedes from the lever. This switches the route of the current flowing through its wires.

The switch used here is an excellent switch, however we would have preferred a tighter dead band of about 2 psi, or possibly a variable dead band for more flexibility. However, many switches with variable dead bands contain mercury, which we wanted to avoid. Plus, we wanted a snap action switch, and they were only available with a fixed dead band version.

This particular brand is hermetically sealed, which was an important consideration because of the switch's proximity to the tank. Also, the snap action switches in this model were potted (sealed). This in itself does not provide an appropriate level of safety, but since the system uses a low current, low voltage signal, we considered the switch to be adequate for our needs and safety concerns. Of course we cannot advise the use of such equipment. Any switches used in a system must meet the standards of your area AHJ.

This switch has a removable plastic see-through cover, which allows observation of the switch, and yet protects the mechanism. We drilled a few holes at the top of the switch case to vent any hydrogen that might accumulate. We had to be careful not to disturb the bourdon tube with the drill bit, as that would compromise the integrity of the tube.

The most appropriate switch for this application would be the Mercoid switch DSH-7341-153-23E. This is the same as the Mercoid switch DS-7341-153-23E except that it comes with an explosion-proof housing. With the addition of the explosion-proof housing, the switch costs almost twice as much.

There are several types of switches that are adequate to perform the required tasks. Do some research in suppliers' catalogs to find exactly what you need. Notice that what are generally considered to be explosion-proof switches tend to be priced on the high side. One reason for this is that not too many people use such switches, so they are not mass produced and therefore not as inexpensive as they could be.

Installing the storage tank peripherals

The pressure tank that we used has four ports: three ¼" NPT ports and one ½" NPT port. We used a ¼" NPT port for the pressure switch and relief valve. The relief valve and pressure switch are connected to the tank by a ¼" NPT, female x female x male T pipe fitting.

- Thread seal and Teflon tape the male end of the T pipe fitting, screw into tank and tighten.

- Thread seal and Teflon tape the male ¼" NPT connection on the pressure switch, screw on and tighten.

- Set the adjustable Brass Pop-Safety Valve for the pressure you want. Even though the valve is adjustable, the distributor has this particular valve preset at around 88 psi. This was fine since the tank is rated at 125 psi and our initial tests were not going to go above pressures of about 60 psi.

- Thread seal and tape the male NPT end of the valve and screw into the T fitting and tighten.

- Attach the wires to the pressure switch from the process controller:
    - Wire #1 in assembly 2 that comes through port 3, to the black Common wire on the pressure switch.

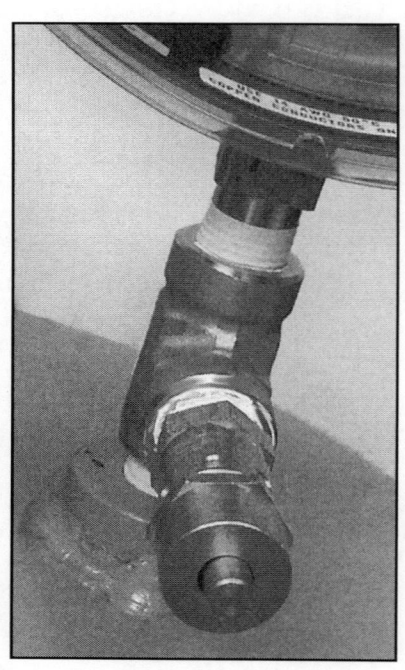

Relief valve and pressure switch connections

Storage Tank and Peripherals

- Wire #1 in assembly 1 that comes through port 2, to the red Low wire on the pressure switch.
- Wire #2 in assembly 1 that comes through port 2, to the blue High wire on the pressure switch.

Pressure switch dead bands

Set the pressure switch to activate at the desired pressures (both low and high). This particular switch model has a fixed dead band of 6 psi. The dead band is the difference between the set activation points (high and low pressure). There is an adjustment knob on the side of the switch. If the high pressure point is set at 66 psi, it will turn the electrolyzer off when the pressure reaches 66 psi. Because the dead band is fixed at 6 psi, when the pressure in the tank falls to 60 psi, the switch will activate again and turn the electrolyzer on.

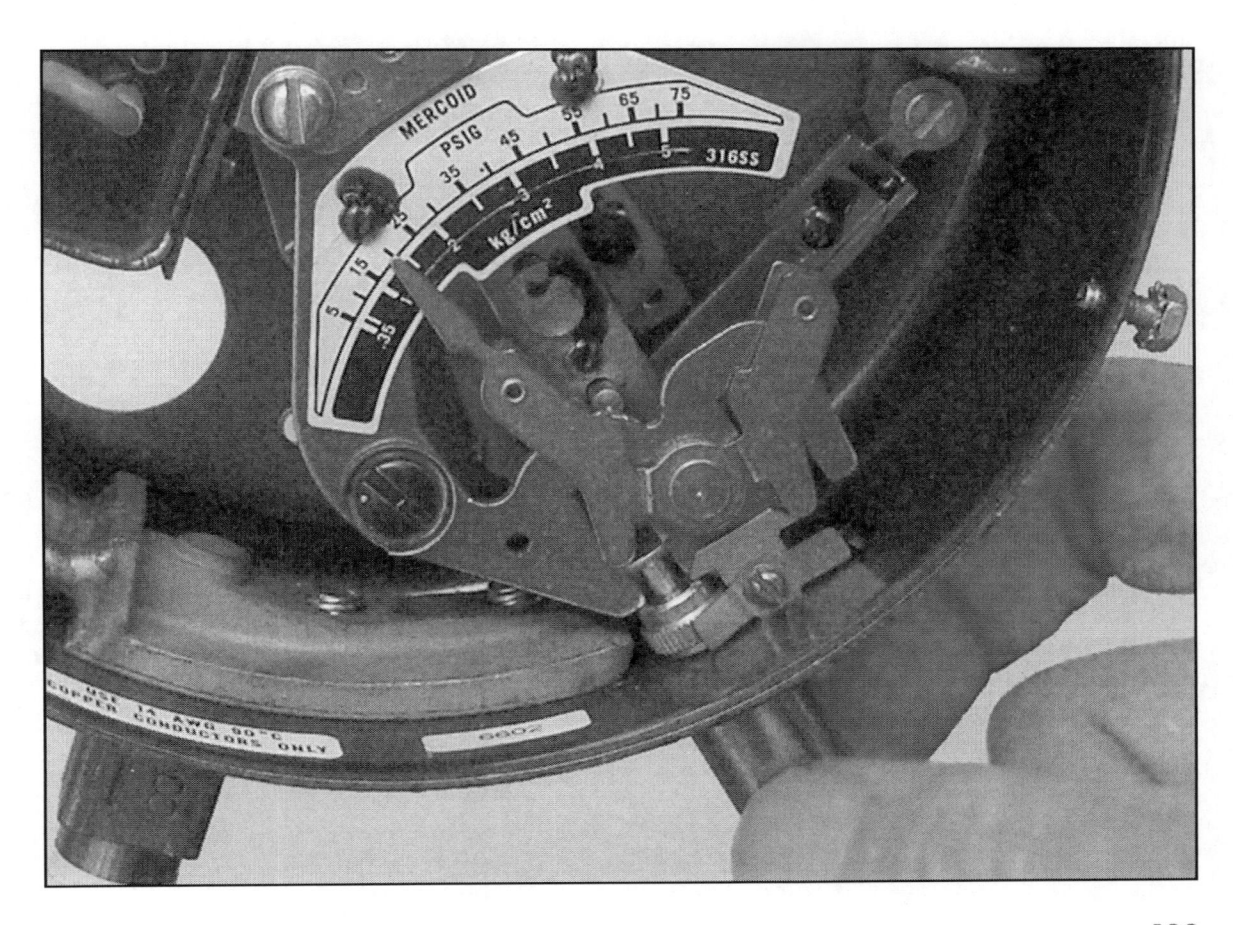

Practical Hydrogen Systems

Regulator installation

- Thread seal and tape the ½" x ¼" hex nipple on both ends, screw into the tank's ½" port and tighten

- Screw the regulator onto the nipple and tighten. For the regulator outlet you can apply any outlet fitting that serves your purpose. We used a ¼" NPT (which fits the regulator outlet) x ⅛" barb hose connection. Be sure to thread seal and Teflon tape.

- The next step is to thread seal and Teflon tape the male NPT end of the pressure gauge and screw that in to the tank and tighten.

Regulator for storage tank

Storage tank and peripherals parts and materials

Description	Supplier	Part #	Quantity
316 SS threaded precision pipe fitting ½" x ¼" pipe, 1²¹⁄₃₂" long, hex nipple, 6600 psi	McMaster-Carr	48805K871	1
316 SS threaded precision pipe fitting ¼" x ¼" pipe size, hex coupling, 6600 psi	McMaster-Carr	48805K69	1
316 SS threaded precision pipe fitting ¼" pipe size, hex head solid plug, 8000 psi	McMaster-Carr	48805K28	1
Portable pancake style pressure tank 6 gallon capacity, 16" diameter, 125 max psi	McMaster-Carr	1862K91	1

Storage Tank and Peripherals

Description	Supplier	Part #	Quantity
Brass multi pressure range pop-safety valve soft-seated, ¼" NPT male, 25-200 psi	McMaster-Carr	50265K23	1
Adjustable brass pop-safety valve, low-temp service ¼" NPT male, 38 to 140 psi range, set @88 psi	McMaster-Carr	9137K47	1
316 SS threaded precision pipe fitting ¼" pipe size, female x female x male T, 5600 psi	McMaster-Carr	48805K571	1
Line Regulator 0-30 psi Operating Pressure, 0-60 psi Gauge	McMaster-Carr	7864A11	1
Pressure switch, hermetically sealed, fixed dead band, snap action, stainless steel.	Dwyer Instruments	Mercoid switch DS7341-153-23E	1
Pressure gauge type 304 SS-case gauge 2% midscale accuracy 2½" dial, ¼" NPT male bottom, 0-100 psi	McMaster-Carr	4066K413	2
Brass single barbed tube fitting barbed x NPT male for ⅛" tube ID, ¼" NPT	McMaster-Carr	50745K36	1

Parts numbers and suppliers have been provided for your convenience; however, suppliers may go out of business and parts numbers may change. All parts listed are available from multiple suppliers.

Practical Hydrogen Systems

Storage tank connections to coalescer

For the simple system without recombiner, flashback arresters and filter, simply connect the coalescer to the tank.

- Connect a ¼" x ¼" NPT stainless steel hex coupling to the stainless steel elbow outlet from the coalescer

- Adjust and test adjustable check valve. The check valve can be set for somewhere from 3 psi to 5 psi. Our particular check valve has an adjustable range of 3 psi to 15 psi. Basically the narrowest range possible is the best. Or, use a non-adjustable check valve set at about 2 psi.

- Insert in tank.

- Connect the stainless steel elbow to check valve.

- Connect the end of the braided hose with a male swivel connection to the elbow on the tank, and connect the other end to the coalescer.

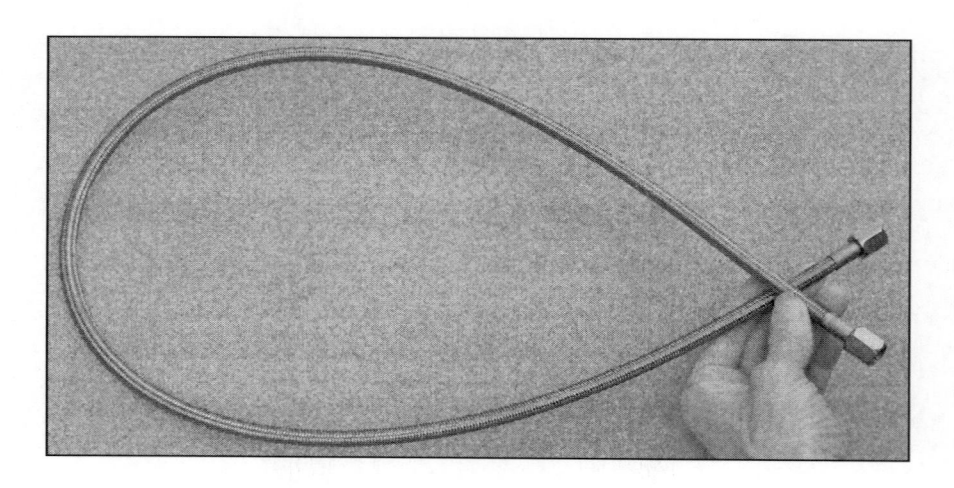

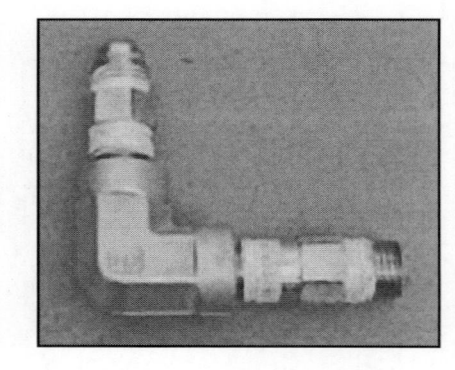

Adjustable check valve assembly

Storage Tank and Peripherals

Other connectors to the tank could be used after installing the check valve in the tank. For instance, a DESO stem and body, and/or other type of quick disconnect will hold the pressure in the line and in the tank. These other connections make it possible to disengage the tank from the system to remove it and/or replace it with another tank.

Materials for storage tank connection

Description	Supplier	Part #	Quantity
PTFE braided hose w/type 304 SS braid swivel male x male fitting, ¼" ID, 3000 psi	McMaster-Carr	4552K422	1 length depending on choice
Extreme-pressure 316 SS threaded pipe fitting ¼" x ¼" pipe size, hex coupling, 6000 max psi	McMaster-Carr	51205K212	1
Precision threaded type 316 SS pipe fitting ¼" pipe size, 90° male elbow, 7500 psi	McMaster-Carr	48805K41	1
Generant adjustable check valve, ¼" male x male, stainless steel	Lindco Inc.	ACV-4P-SS-B-3	1

Parts numbers and suppliers have been provided for your convenience; however, suppliers may go out of business and parts numbers may change. All parts listed are available from multiple suppliers.

Catalytic Recombiner and Subsystem

The catalytic recombiner combines hydrogen and oxygen by catalytic action, which produces heat and water. Its purpose in this system is to further purify the hydrogen stream by removing any oxygen.

This is not a necessary component for electrolyzer systems. Well-designed electrolyzers have their electrodes generously spaced to avoid mixing hydrogen and oxygen gas. When built properly, and with no malfunction present, a good electrolyzer will make extremely pure hydrogen.

There is a minor amount of diffusion through the electrolyte solution. This can be removed to some degree through catalytic recombination, which upgrades the hydrogen derived from the electrolytic process.

Catalytic recombiners can be purchased, or fabricated yourself.

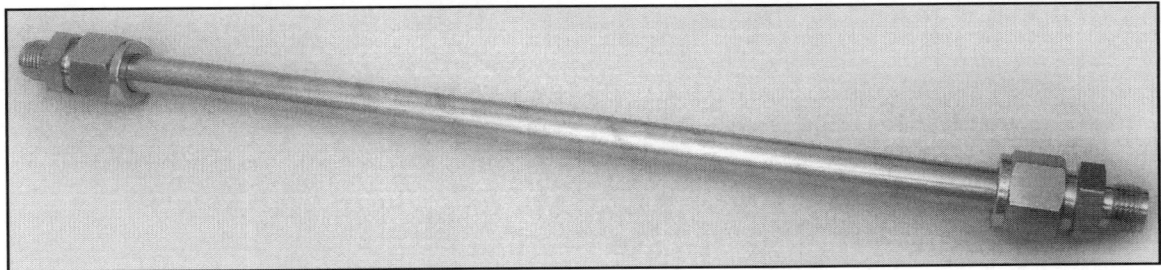

Catalytic recombiner

Building a catalytic recombiner

- ◆ Thoroughly clean and scrub with acetone a ½" OD stainless steel tube. Use a stainless steel brush to clean the inside of the tube, then

Practical Hydrogen Systems

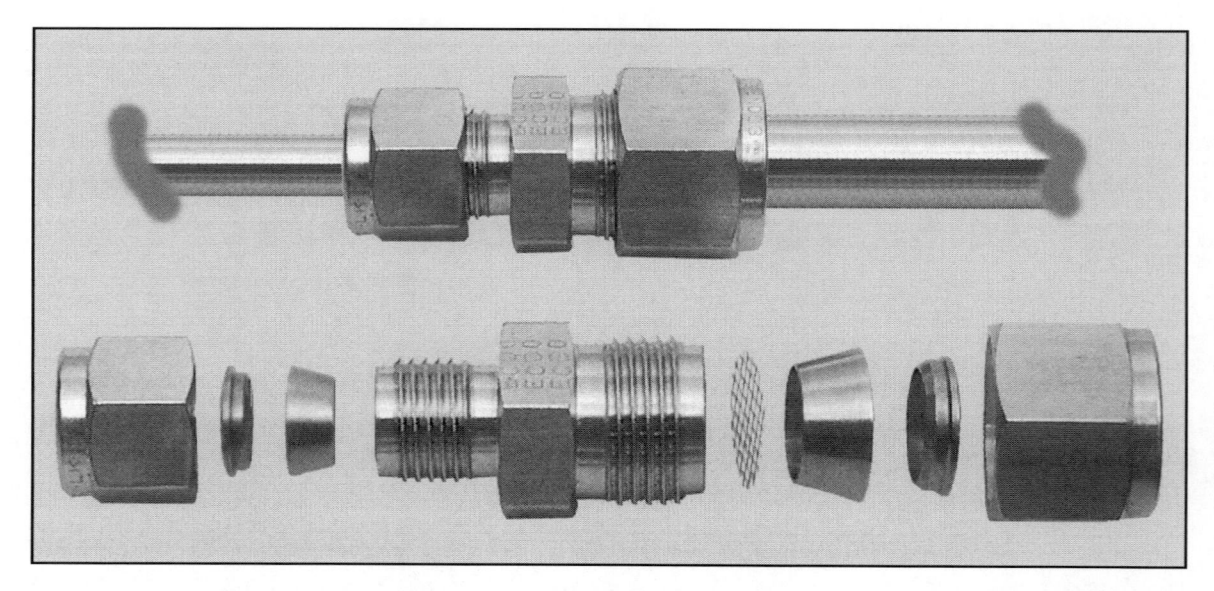

rinse with water. I suggest this for all of the piping, tubing and fittings used for this project. Materials as they come from all suppliers are coated with residues of all sorts. There should be no residual oil or other contaminants in this tube. Do not assume that a component is clean merely because it looks clean.

- ◆ Cut two wire cloth circles to fit inside ½" end of the ½" x ⅜" Yor-Lok fittings. When the tube is put into the fittings, it will hold the screen in place. This screen holds the catalytic beads in the tube, but allows the gas to pass through.

- ◆ Place one of the screens into one of the fittings.

- ◆ Put the tube into the fitting, add the ferrules, slip on the nut and tighten.

- ◆ Pour the platinum catalyst beads into the tube, occasionally tapping and shaking the tube so that the pellets settle. A 25 gram bottle will fill the tube. If it does

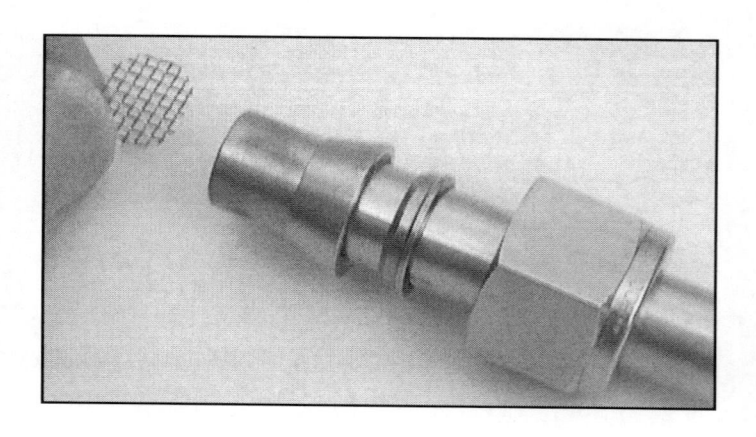

Catalytic Recombiner and Subsystem

not pack right on the first try, gently shake them out and try again. Do not push down on the pellets with an object to squeeze them, because you do not want to break the pellets, or make powder or dust. Work gently, putting them in, tapping and rearranging as you slowly fill the tube.

◆ When the tube is filled, slip the nut for the second fitting over the tube with the two ferrules and put the other piece of screen on top of the tube.

◆ Push the other ½" x ⅜" Yor-Lok fitting on to the ½" tube, slip the ferrules into position and tighten the nut.

The catalytic recombiner is finished.

The catalyst used in the recombiner can be easily poisoned by sulfur compounds, organic solvents, chlorine compounds, and heavy metals such as mercury, lead and cadmium. If you use vinegar in your bubbler system, be absolutely sure that there are no sulfur compounds in it. Any of these components will render the catalyst worthless.

The catalytic recombiner tube must always be placed in a vertical position in the system so that the water drains downward.

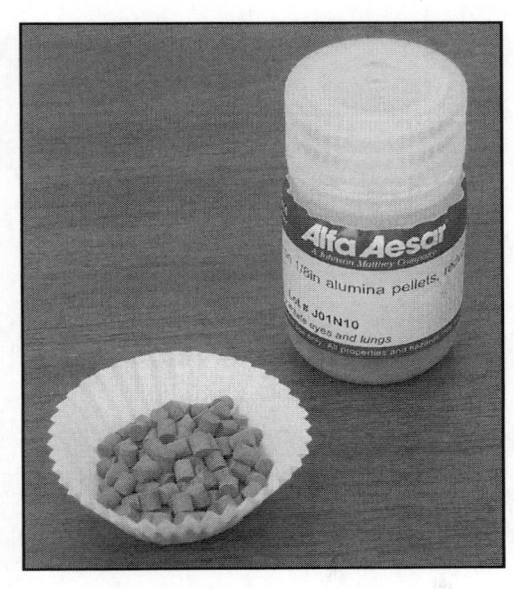

Platinum catalyst beads

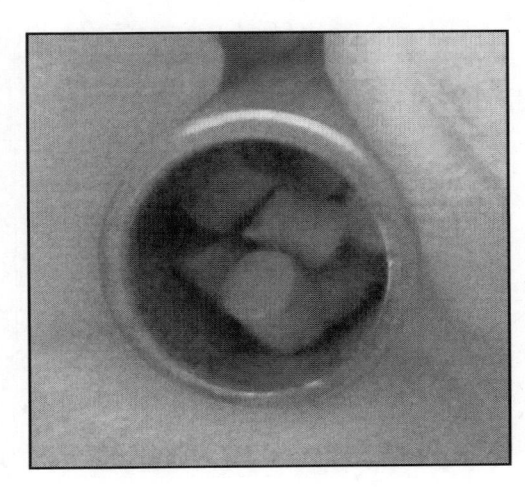

Materials for catalytic recombiner

Description	Supplier	Part #	Quantity
Platinum, 0.5% on 1/8" alumina pellets, reduced, 25 grams.	Alfa Aesar	89106	1
316 SS Yor-Lok compression tube fitting reducing coupling for 1/2" x 3/8" tube OD	McMaster-Carr	5182K368	2
Type 316 SS seamless tubing 1/2" OD, .402" ID, .049" wall, 1' length	McMaster-Carr	89785K247	1
SS milling grade woven wire cloth type 304, 18 x 18 mesh, .015" dia, 12" x 12" sheet	McMaster-Carr	9238T528	1
(Optional) Catalytic recombiner	Specialty Gas Equipment	Series 6210	
(Optional) Catalytic recombiner	Resource Systems Inc.	Series RCP and DBP	

Flashback arresters

When a catalytic recombiner is included in a system, flashback arresters are required on either side of the recombiner. Some flashback arresters can be reset and used again after a flashback occurrence. Other types must be replaced after a flashback occurrence. We used an arrester that could not be reset, but was inexpensive compared to the types that can be reset. These are widely available through welding shop suppliers.

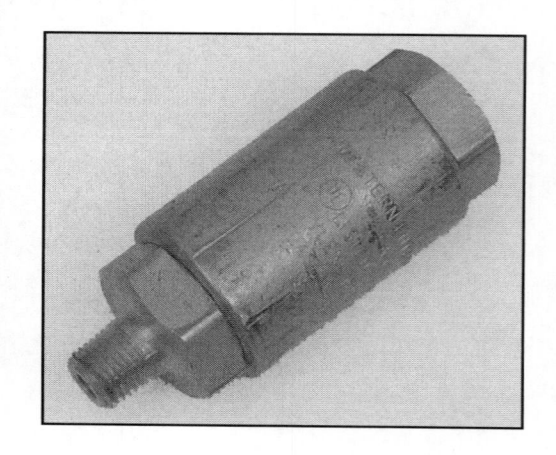

Catalytic Recombiner and Subsystem

This particular arrester has a brass body, as most do, which is not appropriate for long term use in a system with a KOH electrolyte. Stainless steel arresters are preferred, but are more expensive than the brass. Since our system is experimental and is frequently assembled and disassembled, any corrosion problems can be monitored, so we opted for brass.

Source for flashback arrestors

Description	Supplier	Part #	Quantity
Western Enterprises flashback arrester, ¼" NPT female x ¼" male	McMaster-Carr	Western Enterprises part# FA-3	2

Filter

Catalytic recombiners also require a filter, located after the second flashback arrester, to collect the water formed by the catalyst.

Most filters on the market have either aluminum or zinc bodies and polycarbonate bowls. These materials are very reactive and do not stand up well under exposure to KOH. Either a stainless steel filter or coalescer should be used for this purpose.

We used a brass body filter with a polyurethane bowl as a low cost alternative to purchasing another stainless steel bodied filter. This worked well for the experimental purposes, but stainless steel is the appropriate metal for this component in this type of system.

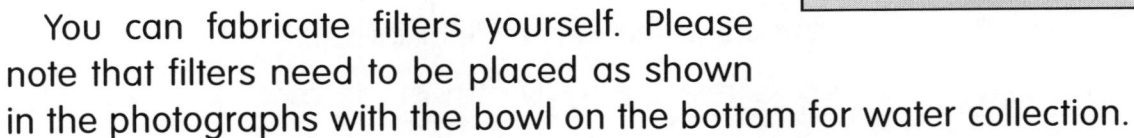

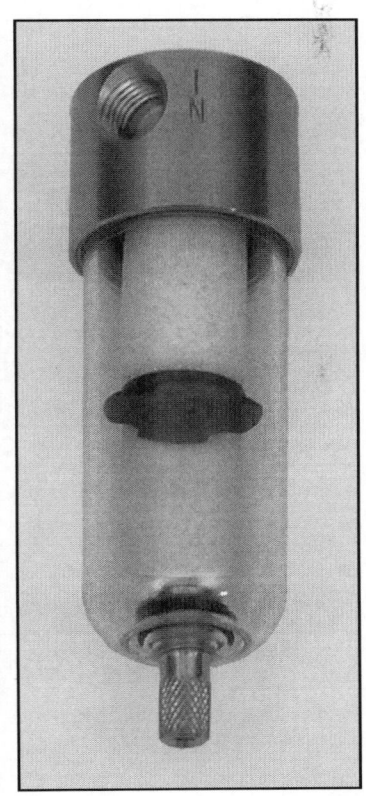

You can fabricate filters yourself. Please note that filters need to be placed as shown in the photographs with the bowl on the bottom for water collection.

Practical Hydrogen Systems

Source for filter

Description	Supplier	Part #	Quantity
Filter with brass body and polyurethane bowl ¼" pipe, 22 scfm max	McMaster-Carr	8287K12	1

Catalytic subsystem assembly

- Screw on the female side of one of the flash arresters to the 90° male exit elbow of the coalescer.
- Thread seal and Teflon tape the male end of the arrester and screw in and tighten to the female end of a Yor-Lok ¼" NPT female x ⅜" fitting.
- Insert a 1½" length of ⅜" tube in the ⅜" tube end of the Yor-Lok fitting and tighten.

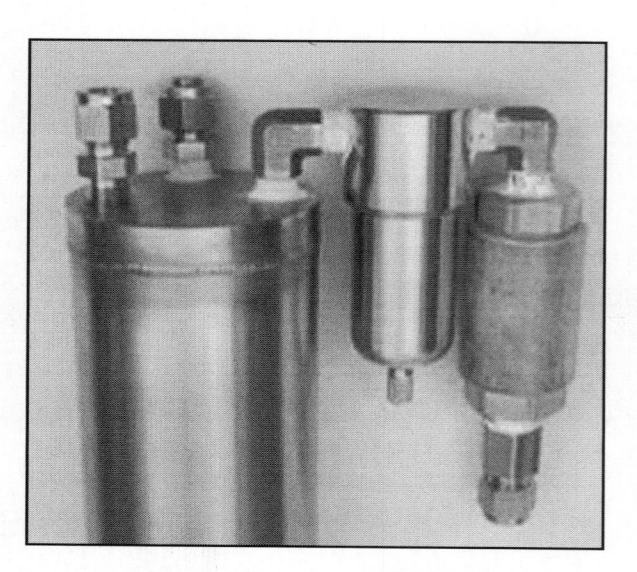

Left to right – bubbler, coalescer, and flashback arrestor

- Attach that tube to the ⅜" Yor-Lok outlet at one end of the recombiner and tighten.
- Attach a 1½" piece of ⅜" tubing to the other end of the catalytic recombiner and tighten.
- Attach the other end of the ⅜" tubing to the tubing end of a Yor-Lok ¼" NPT male x ⅜" tube fitting.
- Thread seal and tape the ¼" male end of the elbow and insert and screw in and tighten into the female filter inlet port.
- Thread seal, Teflon tape the male end of the other flash arrester. Screw and tighten this into the female outlet port of the filter.

Catalytic Recombiner and Subsystem

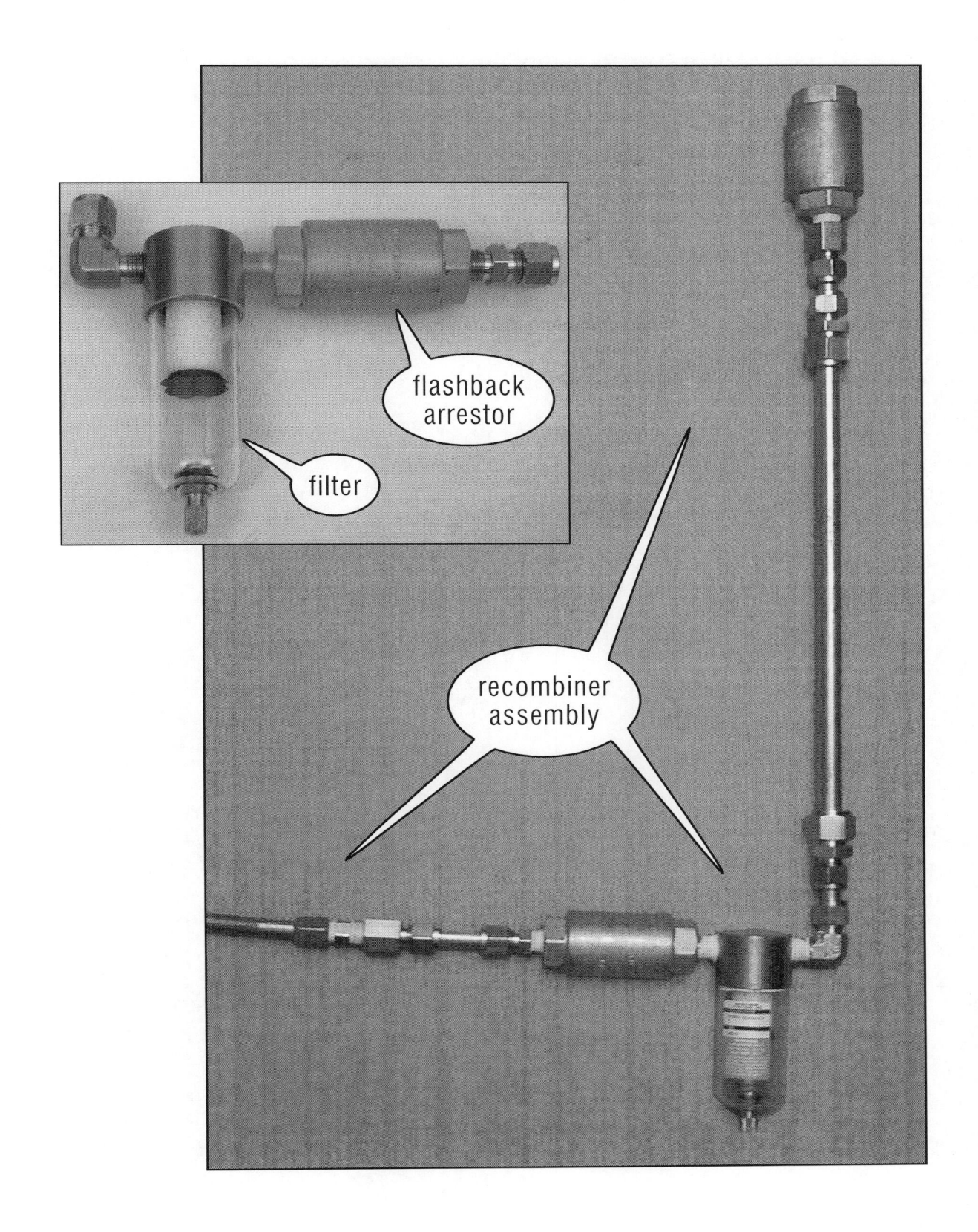

Practical Hydrogen Systems

- Thread seal and Teflon tape the male end of the ¼" NPT male x ⅜" tube Yor-Lok fitting, and screw into female side of arrester.
- Insert 1½" length of ⅜" tube into tube end of this fitting.
- Attach the tube end of a Yor-Lok ¼" female x ⅜" tube fitting to the ⅜" tube and tighten.
- The ¼" NPT female end of this fitting is then attached to the flexible metal hose with swivel male fitting after you thread seal and tape. The hose part number is in the storage tank connections section on page 187**.

Catalytic Recombiner and Subsystem

Fittings for catalytic recombiner subsystem

Description	Supplier	Part #	Quantity
316 SS Yor-Lok compression tube fitting female straight adapter for 3/8" tube OD, 1/4" NPT	McMaster-Carr	5182K266	2
316 SS Yor-Lok compression tube fitting 90° elbow, tube x male for 3/8" tube OD, 1/4" NPT	McMaster-Carr	5182K156	1
316 SS Yor-Lok compression tube fitting male straight adapter for 3/8" tube OD, 1/4" NPT	McMaster-Carr	5182K119	1
Type 316 SS seamless tubing 3/8" OD, .305" id, .035" wall, 3 pieces, each 1½" long	McMaster-Carr	89785K837	1 6' length

Parts numbers and suppliers have been provided for your convenience; however, suppliers may go out of business and parts numbers may change. All parts listed are available from multiple suppliers.

Process Controller

Our process controller was designed for a very low current, low voltage signal in a hazardous atmosphere.

Below is an overview diagram of the electrical system. There are two sensors in the system, the pressure switch with a bourdon tube sensor on the storage tank, and a fill level switch which controls the fluid level of the electrolyte. The fill level switch can be designed for a

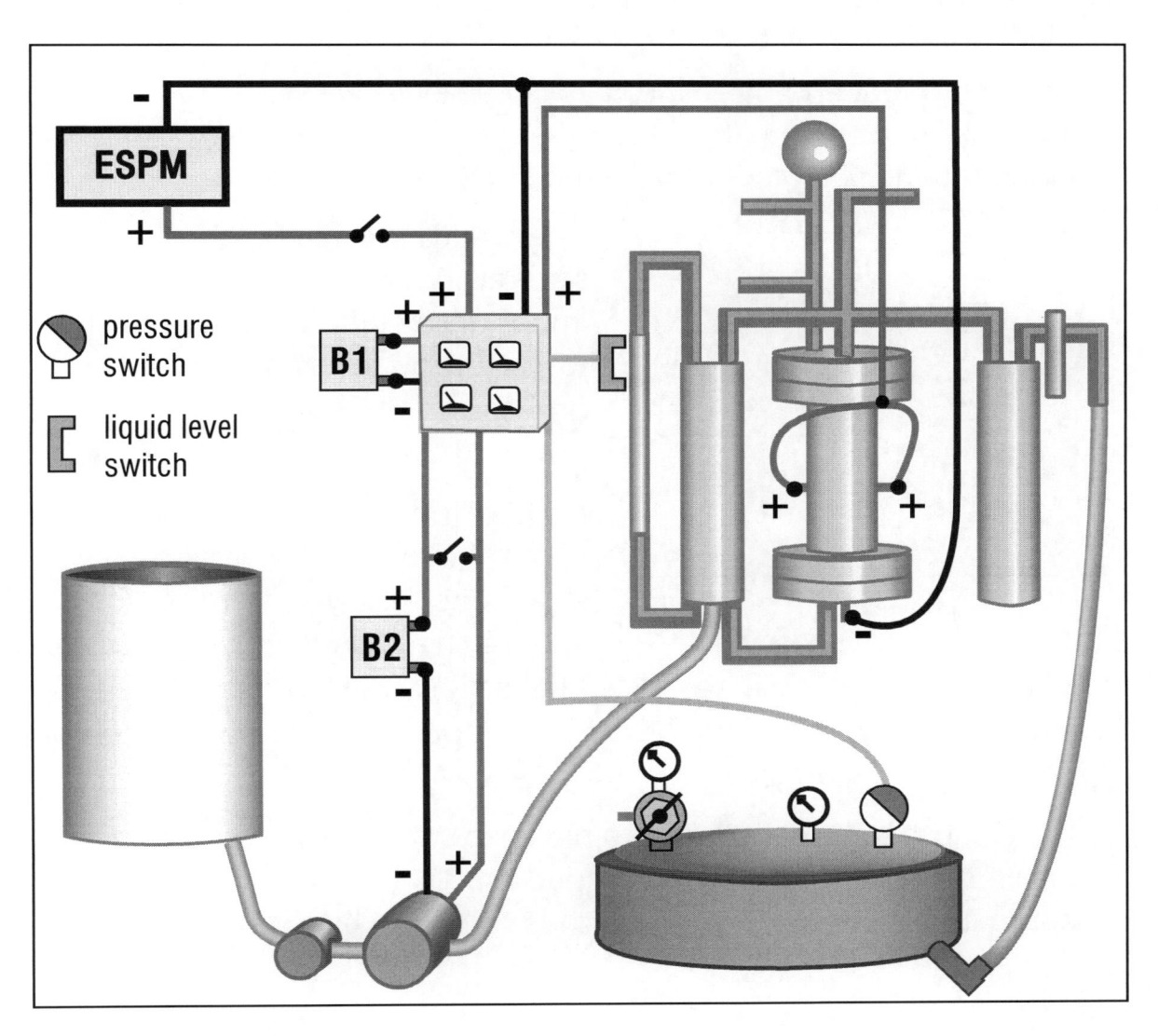

Practical Hydrogen Systems

number of sensor options, including infrared and visible light photonic, with or without fiber optics; capacitive; inductive; ultrasonic; and others.

The system utilizes triple witch switches for the electrolyte level sensors and for the storage tank pressure sensor. These consist of a combination of three switches – first, the sensor; second, the SCR switch and third, a relay.

When a sensor detects a predetermined state, a signal is sent to the SCR. This allows current to flow through to the relay, energizing the relay coil.

The signal that energizes the coil also de-energizes the coil. This is where the witch part comes in. The signal turns the coil on and off at the same time.

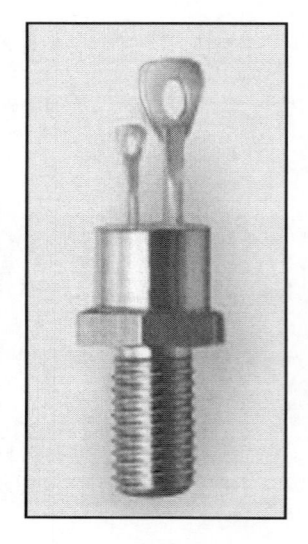

Silicon controlled rectifier (SCR)

Relay switch

The relay is constructed to perform several functions at once, the most important being to supply power or to cut off power to the electrolyzer, and to supply power or to cut off power to the electrolyte fill pump.

This triple witch configuration allows the power circuitry to be several steps removed from the atmosphere the signal must be generated in, which makes it safer in a hazardous atmosphere. There are many other ways this could have been done, but this is the method we chose for this project.

The process controller also has a pressure switch for the process control box with attendant power supply; and an LCD pressure readout for the box with its own power supply, which is controlled by a magnetic switch.

Process Controller

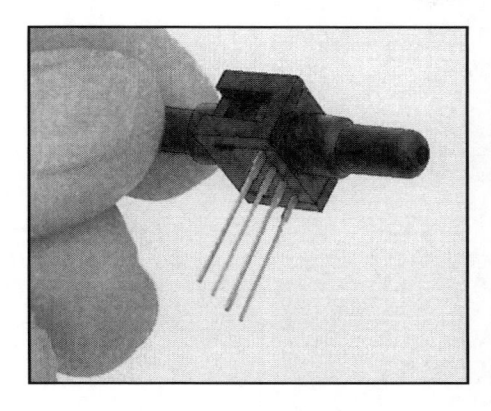

Pressure sensor for process control box

The hydrogen tank pressure sensor switch controls according to preset values, the cutoff and startup points for the operation of the electrolyzer. When the pressure reaches the preset cutoff pressure, a signal is sent to turn the electrolyzer off. When it falls to the preset startup pressure, it will turn the electrolyzer on again.

The particular sensor/switch that we chose for this project has a bourdon tube to sense pressure and mechanically switch the signal current. It is an electromechanical switch.

The switch itself is installed on the top of the storage tank. As the tank fills and pressure builds, the calibrated metallic bourdon tube bends. When the cutoff pressure level is reached, a mechanical lever is thrown and switches the current off. When the tank pressure falls to the startup pressure level, the mechanical arm is activated and the electrolyzer turns on again.

The current that the pressure switch turns on or off is provided by the voltage divider, current reducing circuit in the controller which consists of three resistors.

The resistors lower the voltage and current being sent to the external circuit in the hazardous atmosphere to about one volt and about 20 milliamps. This current in the external circuit is controlled by the pressure switch on the pressure tank. The current flows through one wire or another depending on the position of the switch. As the current is allowed to flow it triggers the SCR (silicon controlled rectifier) which in turn allows a higher voltage and current to flow through it to the relay. The relay, when activated, will switch the solar panels or other power supply to the electrolyzer on or off.

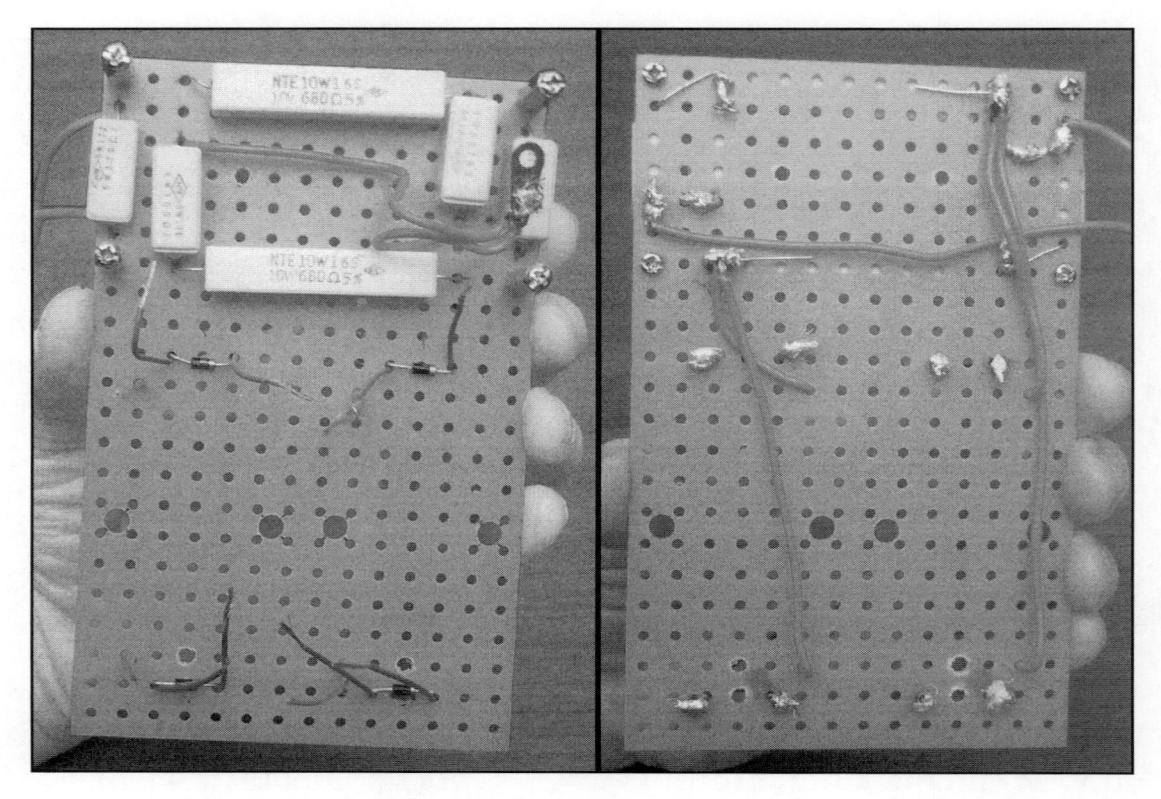

Top and bottom of circuit platform containing voltage divider resistors

This particular circuit in the controller consists of one voltage divider/current limiter, two SCRs and one relay.

The pump/fill level circuit also consists of one voltage divider/current limiter, two SCRs and one relay. It works the same way as the pressure switch circuit.

The process controller also contains an internal pressure sensor that is powered through a LM317 voltage regulator circuit and a LCD readout powered by an internal 9 volt battery with a magnetic reed on/off switch. This switch is activated by a magnet applied to the reed switch from the outside of the box. The magnet can be held in position with Velcro or can be placed into a slide holder. To remove power to the LCD, you simply remove the magnet. This was not a necessary

Process Controller

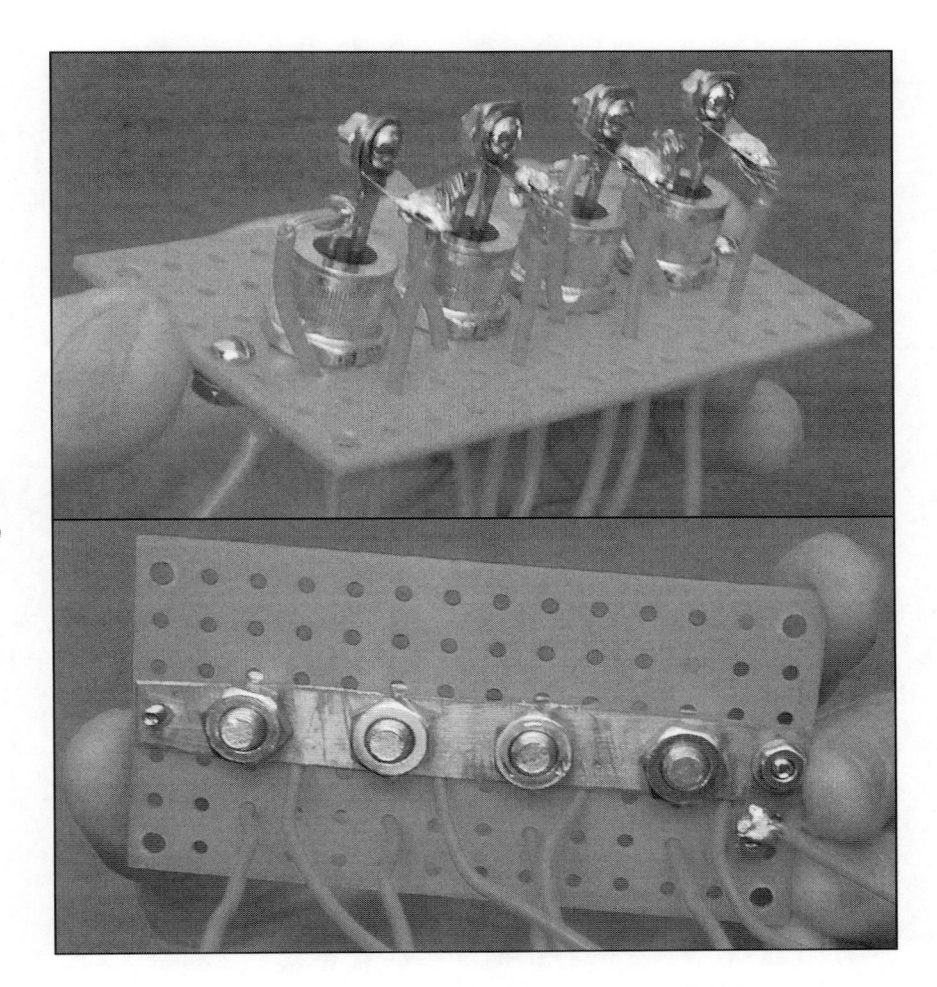

Top and bottom of SCR platform

part of the design, but we wanted to experiment with some different approaches to activating particular circuits and this was one result. It is an excellent way to control circuits in an enclosed pressurized container that you do not want to compromise.

Practical Hydrogen Systems

Process controller subassemblies

The hydrogen process controller is housed in a pressurized box.

It contains two circuit platforms which hold the components for several assemblies critical to the function of the system.

◆ Platform One contains a pressure sensor for the box and attendant power supply regulator with a digital read out.

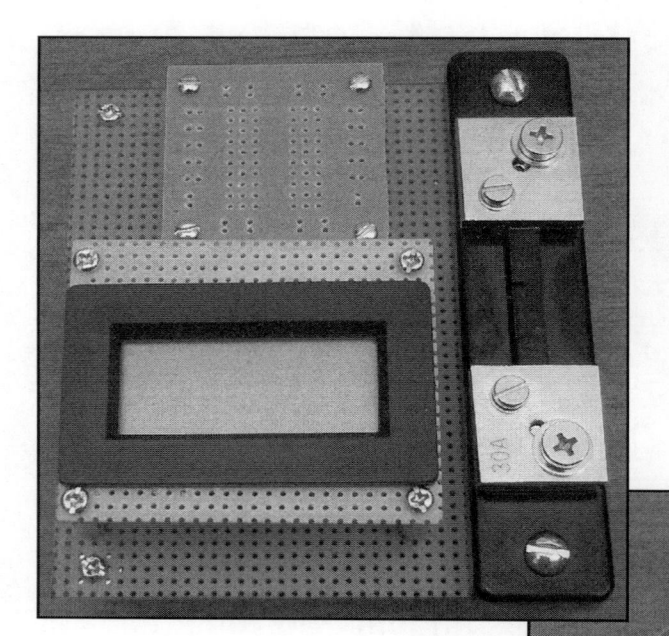

Partial assembly, Platform One. Includes curcuit board for voltage regulator/pressure sensor.

Completed circuit board platforms containing process control box pressure sensor units and shunt (Platform One).

204

Process Controller

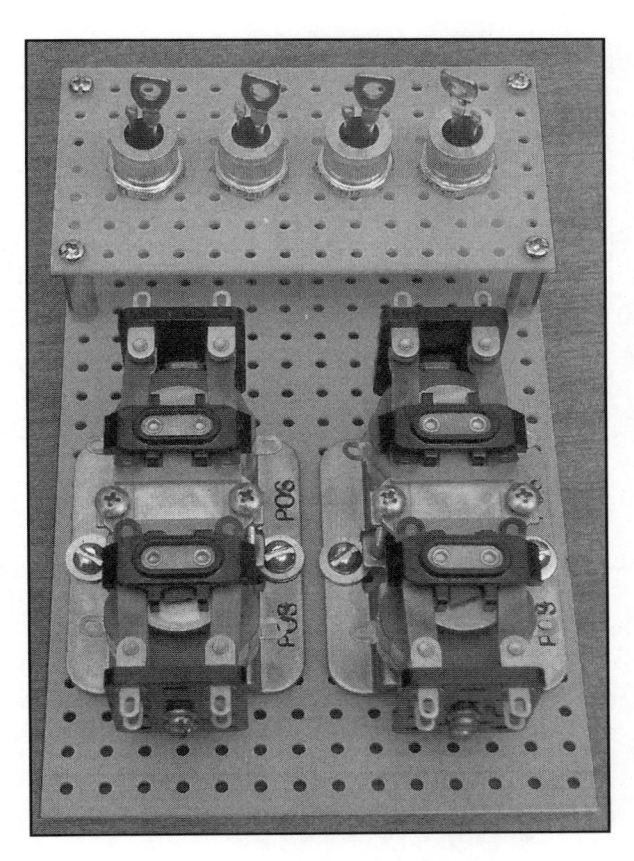

◆ Platform Two contains a bank of SCRs (silicon controlled rectifiers), relays and voltage dividers.

Relay switch assembly, Platform Two

A flat plastic insert is part of the meter assembly which consists of four analog meters for observing current and voltage for different functions of the system.

205

The controller has five ports, two for pressure and three for wire. The box also contains a nine volt battery supply with a magnetic switch for a digital voltmeter, and a variety of terminal block connectors.

This controller was designed for ease of fabrication, relatively low cost, with a minimum of components. For the above reasons the system is not by any means optimal, but does do the job adequately for the purpose of learning the basics in regard to the design and building of an experimental hydrogen system.

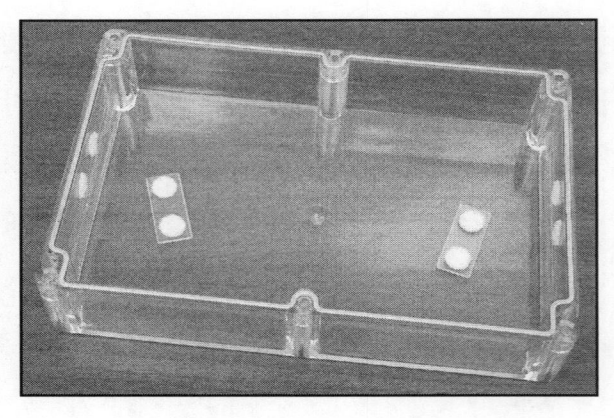

Box front with analog meter assembly tab holders

Stand-offs and terminal blocks in place, pressure sensor port in place, four holes in bottom for ports

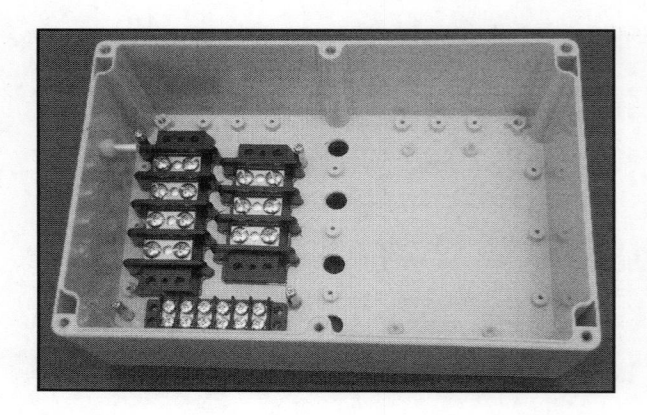

Process controller enclosure

The enclosure is a NEMA type 4x, which is designed to be used either indoors or outdoors in a nonhazardous location, and which allows some protection from dirt particles, general precipitation, ice, splashing and hosed water, and general contact for the components within it.

The design of any electrical control unit that will operate in a hazardous location must address the issue of safety, and meet your AHJ's regulation requirements for its fabrication and fitness for its operating environment.

NEMA (National Electrical Manufacturers Association) provides enclosure classifications based on environment of use. For hazardous environments the NEMA recommendation for a hydrogen generating and processing systems specifies a Type 7 enclosure which is designed to stand up against an internal explosion and contain it without igniting a dangerous hydrogen/air mixture outside of the enclosure.

Explosion-proof enclosures

Although the term used for this type of enclosure is "explosion-proof," it does not mean that the atmosphere inside the enclosure won't ignite and explode. It simply means that if there is an internal explosion, the internal explosion will not escape the enclosure and ignite an external explosion.

The guidelines for the manufacture of these enclosures include that they must be strong enough to withstand an internal explosion and they must be flame tight. This requires enclosures to be made of metal with well machined flanges and tight tolerances so that hot gas will cool as it exits across the extended flange faces into the surrounding atmosphere. A flange interface can either be threaded and or flat. Needless to say NEMA Type 7 enclosures are more expensive because of the duty they are required to perform.

Pressurized enclosures

We chose a NEMA 4x enclosure and pressurized it enough so that if a leak did occur, the air inside the enclosure would move outward, and not allow hazardous gases to enter. Since the unit always had an operator in attendance to observe critical pressure changes with a dedicated pressure sensor, this was a viable alternative for short term testing of the system. The operator shuts down the system as soon as a problem is observed.

Purged and pressurized enclosures

For hazardous areas a purged and pressurized, or pressurized system and/or a NEMA 7 enclosure is appropriate. Purged and pressurized systems are evacuated and then refilled with an inert gas such as nitrogen. A pressurized system is simply filled with air at a pressure higher than the external air pressure.

The NFPA (National Fire Protection Association) publication # 496 details the requirements for design and operation of purged and pressurized systems for hazardous areas.

Placement of electrical components

Electronic control enclosures as well as other electrical equipment are best placed at a lower level, and away from gas generating units, process lines and storage containers. This is because hydrogen rises rapidly away from the ground and normally disperses quite quickly. Such placement is, however, not a substitute for appropriate enclosures since a slight breeze or draft can blow a hydrogen emission sideways and/or downwards, even when adequate ventilation has been provided for the gas to escape upwards and not collect.

The control unit can also be isolated from the hazardous area with an appropriate barrier according to code. To understand the appropriate placement of electronic and electrical equipment in hazardous locations, review The NFPA's National Electrical Code, referred to as NFPA 70. Articles 500-516 are particularly relevant.

Specific electrical hazards

Relays such as those used in this particular system can be extremely dangerous because the contact points can spark and very easily ignite a fuel rich atmosphere, which will cause an explosion. Keep the controller well away from the rest of the system, or house it in a pressurized unit with alarm or alert shutdown controls, or in a flanged explosion-proof enclosure.

All wiring and electrical components used in any control system must be rated for the current and voltage carried. If they are not, there is a risk of overheating components and/or starting a fire on

Process Controller

wire insulation, etc. All electrical connections whether mechanical or soldered need to be done well so that shorts do not occur between wires, terminals or soldered joints.

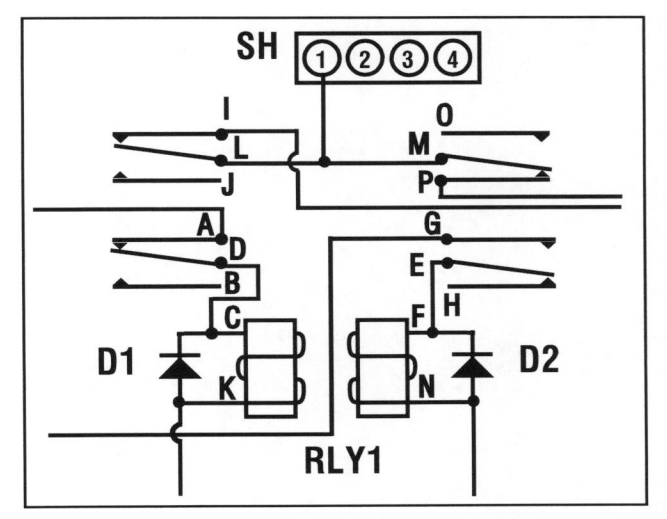

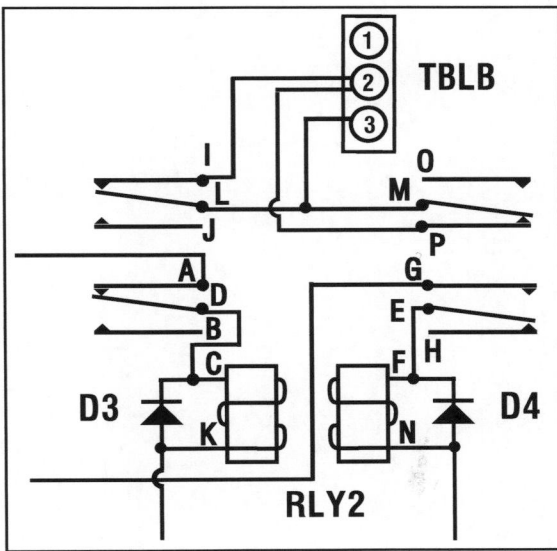

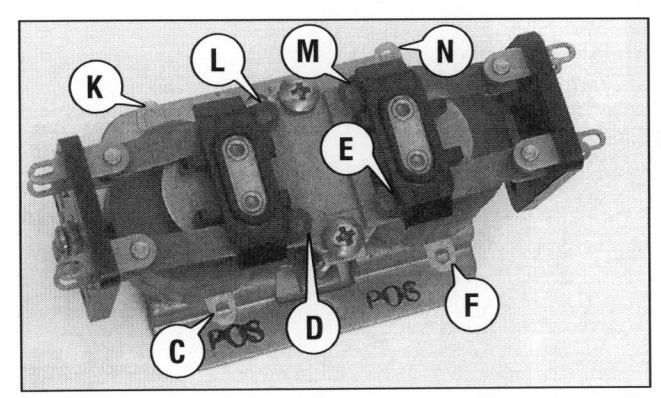

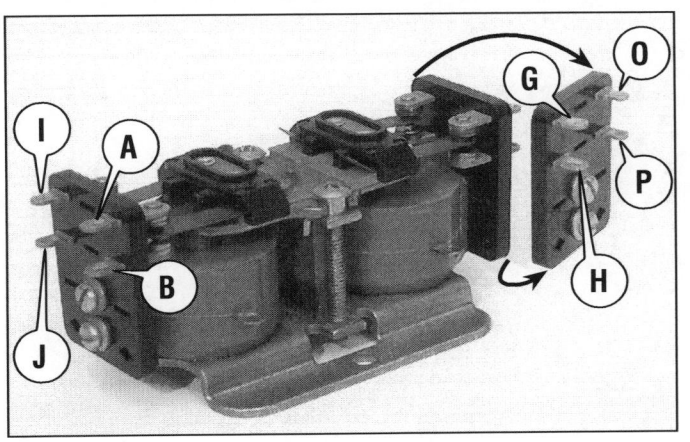

Relay Schematics

SH - Shunt
TBL - Terminal block connector
D - Diode
RLY - Relay

209

Process Controller Schematic Legend

AAM - Analog ammeter
AMAM - Analog milliammeter
A - Anode
AVM - Analog voltmeter
B - Battery
DPVM - Digital panel voltmeter
D - Diode
ESPM - Electrolyzer specific photovoltaic module
E - Electrolyzer
F - Fuse
G - Gate
K - Cathode
R - Resistor
M - Magnet

RLY - Relay
SCR - Silicon controlled rectifier
SH - Shunt
PRSW - Pressure Switch
PU - Pump
PUSW - Pump switch
RSW - Reed switch
TBL - Terminal block connectors
TSW - Toggle switch
VR - Voltage regulator
W - Wire

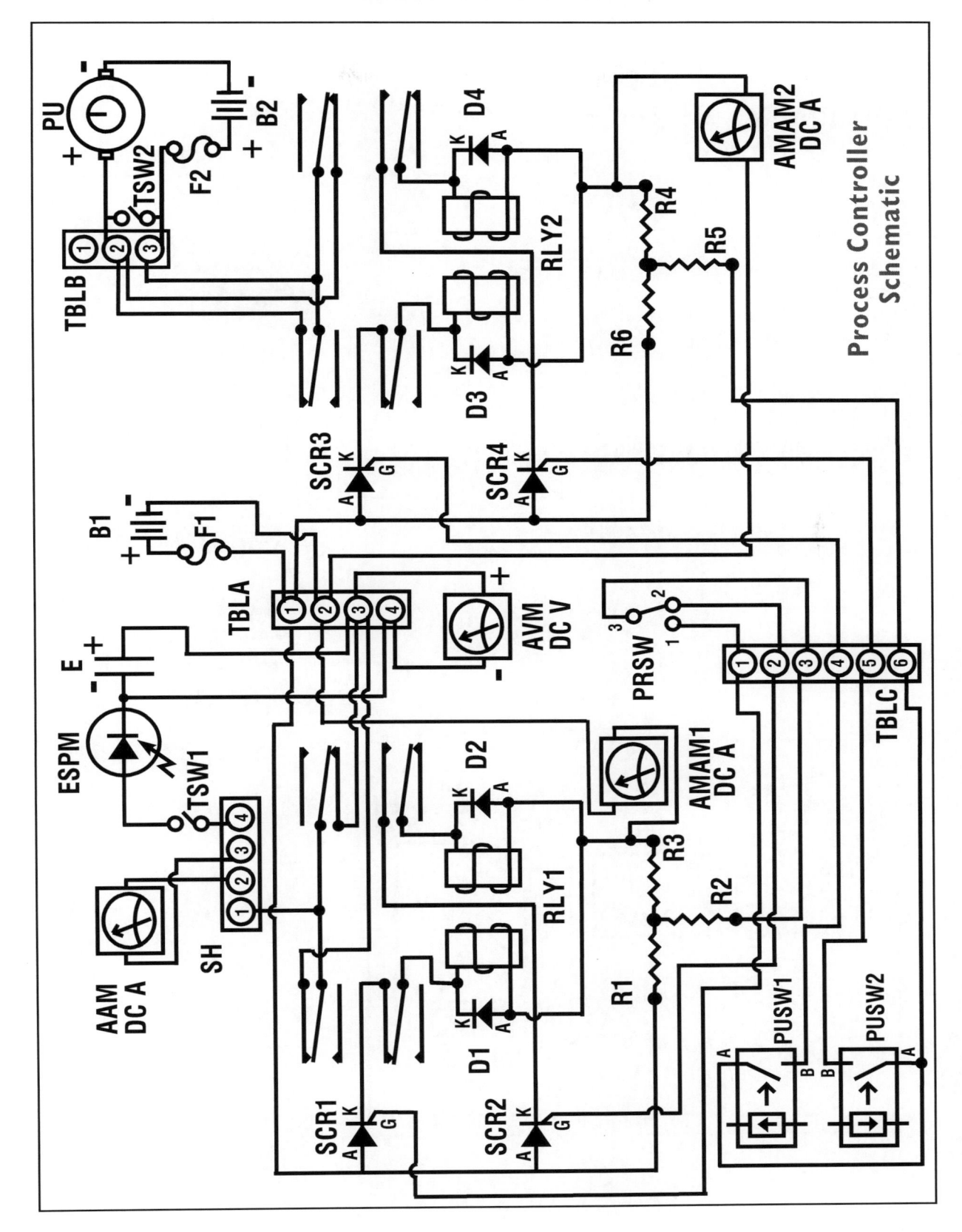

Process Controller

Process Controller Schematic

Practical Hydrogen Systems

Voltage Regulator Schematics

RSW - Reed switch
F - Fuse
M - Magnet
TBL - Terminal block connector
VR - Voltage regulator
R - Resistor
PRS - Pressure sensor
B - Battery
DPVM - Digital panel volt meter

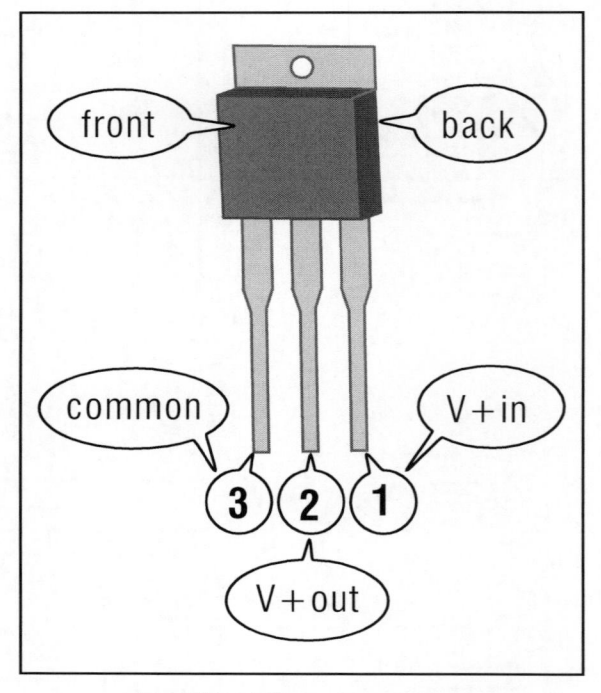

LM317T Voltage regulator

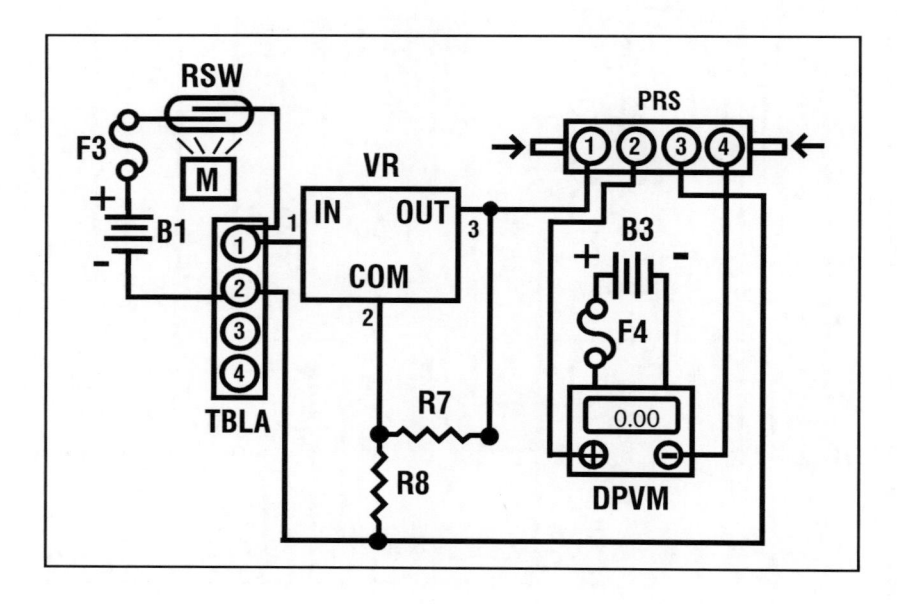

Process Controller

Parts description per legends

AAM - Analog 30 amp DC with shunt

AMAM 1, 2 - Analog 0-50 DC milliammeter

AVM - Analog 0-15 volt DC voltmeter

B1 - 12 volt rechargeable battery

B2 - 12 volt rechargeable battery

B3 - 9 volt rechargeable battery

D1, D2, D3, D4 - Diode 1N4001, 1 amp

DPVM - LCD digital panel volt meter

ESPM - low voltage, high current solar panel

F1 - Fuse, 5 amp

F2 - Fuse, 20 amp

F3, F4 - Fuse 2 amp

M - Magnet

PRS - Differential pressure sensor, 0-5 psi

PRSW - Pressure switch, hermetically sealed, fixed dead-band, stainless steel, snap action switch

PUSW - Pump switch

PU - 12 volt DC, 10 amp, demand pump, 60 psi max.

RLY 1, RLY 2 - Open frame 4 PDT 12 volt DC latching relay, 20 or more amp capacity.

R1, R6 - 100 ohm 5W power resistors

R2, R5 - 120 ohm 5W power resistors

R3, R4 - 680 ohm 10W power resistors

R7 - 220 ohm ½W resistor

R8 - 1600 ohm ½W resistor

SCR 1, SCR2, SCR3, SCR4 - 35 amp silicon controlled rectifier, 600 PRV

SH - Shunt, 30 amp

RSW - Magnetic reed switch, SPST, NO, 1 amp

TBL A - 4 position 30 amp or more DC terminal block

TBL B - 3 position 30 amp or more DC terminal block

TBL C - 6 position 10-20 amp DC terminal block

TSW 1 - Electrolyzer power supply switch

TSW 2 - Manual pump switch

VR - LM 317T Voltage regulator

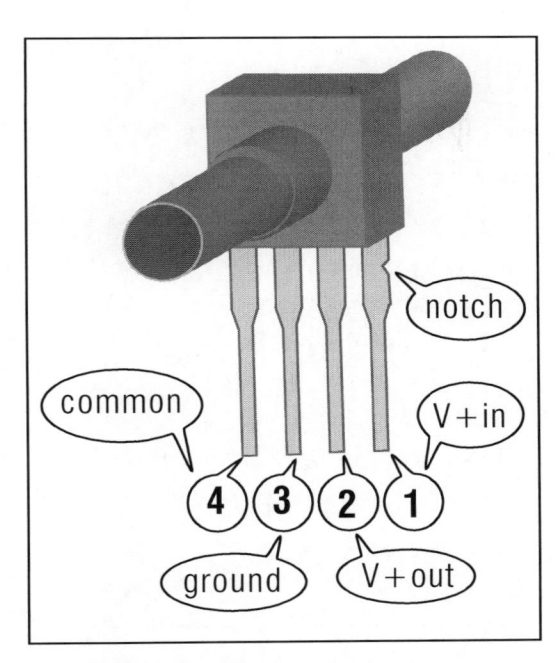

Differential pressure sensor

213

Tools for the process controller

TOOL	PART #	SUPPLIER
Soldering iron, solder		Electronics supplier
Drill, either hand or drill press		
Multimeter		Electronics supplier
Hand-held vacuum/pressure pump zinc alloy head, 25" hg maximum vacuum, 15 psi	9963K21	McMaster-Carr
Plastic cutting tool		Hardware store
Drill and tap for ¼" NPT ports in box		McMaster-Carr, hardware store
Small drill for pressure sensor port hole in box		McMaster-Carr, MicroMark, hardware store

Parts numbers and suppliers have been provided for your convenience; however, suppliers may go out of business and parts numbers may change. All parts listed are available from multiple suppliers.

Materials for the process controller

DESCRIPTION	SUPPLIER	PART #	QUANTITY
Clear silicon rubber caulking	Hardware store		
Plastic tubing, variety of sizes	Hardware store		
Wire - red and black zip cord	All Electronics, automotive or hardware store	WRB-10 (10 ga..), or WRB12 (12 ga..)	
Electrical tape, liquid electrical tape, liquid rubber or shrink tube	Hardware or electronics supplier		
Epoxy glue	Hardware store		

Process Controller

Materials for the process controller

DESCRIPTION	SUPPLIER	PART #	QUANTITY
RLY 1, RLY 2 latching open frame relay; 12 vdc coil - 4PDT; 10 amp contacts; 20 amps total capacity	Surplus Sales of Nebraska	(KO) 110636	2
TSW 1, TSW 2 SPDT 35 amp toggle switch	Surplus Center	11-2157	2
RSW - Switch, magnetic reed, SPST, NO, 1 amp	Jameco	164081	1
Fuse holders	Electronics supplier		2
SCR 1, SCR 2, SCR 3, SCR 4 Silicon controlled rectifiers, 35 amp, 600PRV, C38M	Jameco	14859CR	4
(Optional) Diodes, 6 amp, 50PRV	Jameco	177754	10
(Optional) Zener diodes	Ocean State Electronics	1N4728 through 1N4745	18
Phenolic perf board (example) Size of board or boards needed will depend on your design	Ocean State Electronics	22-518	1
Six position barrier strip	Radio Shack, electronics supplier	274-659	1
Assortment of connectors: ring, spade, easy connect, etc.	Electronics, hardware, or automotive store		
AVM - DC voltage panel meter 0-15v	Radio Shack, electronics supplier	22-410	1

Practical Hydrogen Systems

Materials for the process controller

DESCRIPTION	SUPPLIER	PART #	QUANTITY
DPVM - Digital LCD panel meter	Jameco, electronics supplier	108388CR	1
AAM - DC ammeter - 30 amp. switchboard meter with shunt	Marlin P. Jones, electronics supplier	8733ME	
TBL A - Four position, DC terminal block, barrier strip, 30 amp or more	McMaster-Carr	7527K24	1
TBL B - Three position, DC terminal block, barrier strip, 30 amp or more	McMaster-Carr	7527K23	1
TBL C - Six position, 1020 amp DC terminal block	Radio Shack	274-659	
Metal spacers/standoffs, 11/16" length	Radio Shack	276-195	3 sets
Assorted spacers/standoffs, machine screws and nuts, 4-40	Ocean State Electronics, electronics supplier		
Cable straps (optional)	Hardware store, electronics supplier		
Zip cord and hook up wire, 10 ga., and assortment - 18 ga., 20 ga., 22 ga., stranded	Electronics supplier, hardware store		
Assorted plain perforated board pieces without copper cladding	Electronics supplier		
Nema 4x sealed polycarbonate box with clear cover	Allied Electronics	#736-2041	1

Process Controller

Materials for the process controller

DESCRIPTION	SUPPLIER	PART #	QUANTITY
Brass hose fittings, barb x male pipe for ½" hose ID, ¼" pipe	McMaster-Carr	#5346K22	3
Miniature chrome plated brass ball valve, screw driver handle, ¼" NPT male x ¼" NPT female	McMaster Carr	#4912K97	1
Barb & male pipe for ⅛" hose ID, ¼" pipe male	McMaster-Carr	#5346K61	1
Plexiglas or plastic sheet for meter backing	Hardware store		
PRS - Pressure sensor, gauge or differential	Allied Electronics	#642-2265 for gauge, or 642-2266 for differential	1
Diode, silicon rectifier, 1N4001, 1 amp.	Jameco, electronics supplier	14859CR	2-10
AMAM 1, AMAM 2 - Analog 0-50 DC milliammeter	Marlin P. Jones	14479ME	2
R1, R6 - 100 ohm 5W power resistors	Ocean State Electronics	R5-100	2
R2, R5 - 120 ohm 5W power resistors	Ocean State Electronics	R5-120	2
R3, R4 - 680 ohm 10W power resistors	Ocean State Electronics	R10-680	2
R7 - 220 ohm ½W resistor	Ocean State Electronics	220 ohm, ½W resistor	1

Practical Hydrogen Systems

Materials for the process controller

DESCRIPTION	SUPPLIER	PART #	QUANTITY
R8 - 1600 ohm ½W resistor	Ocean State Electronics	1600 ohm ½W resistor	1
M - Magnet	McMaster-Carr	See catalog	1
VR - LM 317T voltage regulator	Jameco, electronics supplier	23579CE	1
B1, B2 - Battery, 12V, rechargeable	Source depending on type of battery		2
B3 - 9V rechargeable battery or regular alkaline battery	Electronics supplier		1
D1, D2, D3, D4 - Diode 1N4001, 1 amp	Jameco, electronics supplier	35975CR	4
IC PC Board	Radio Shack	276-159	1
8 pin, solder type DIP sockets for holding LM317 and pressure sensor pins	Radio Shack, electronics supplier	276-1995	2
F1 - Fuse, 5 amp	Radio Shack, electronics supplier	270-1011	1
F2 - Fuse, 20 amp (can use other values depending on pump)	Radio Shack, electronics supplier	270-1074	1
F3, F4 - Fuse 2 amp	Radio Shack, electronics supplier	270-1007	2
Washers	Hardware store		as needed

Process controller external wire assembly connections

Wire assembly 1 through port 2

WIRE	CONNECT TO
1	PRSW 2 (Red wire (low) on Mercoid/Dwyer DS-7341-153-23E)
2	PRSW 1 (Blue wire (high) on Mercoid/Dwyer DS-7341-153-23E)
3	PUSW 1B
4	PUSW 2B
5	B1+
6	B1-

Wire assembly 2 through port 3

WIRE	CONNECT TO
1	PRSW 3 (Black wire (common) on Mercoid/Dwyer DS 7341-153-23E
2	ESPM +
3	B2+
4	PU+

Wire assembly 3 through port 4

WIRE	CONNECT TO
1	E+
2	nothing (free floater)
3	E-
4	PUSW 1A, PUSW 2A

Internal wire guide (process controller)

TBLC

POSITION	CONNECT TO
1	SCR1G
2	SCR2G
3	R2
4	SCR3G
5	SCR4G
6	R5

TBLA

POSITION	CONNECT TO
1	SH1
	SCR 1A, 2A, 3A, 4A
	VR1
2	AMAM 1 and 2
	R8
3	RLY 1, I and P
	AVM 1
4	AVM 2

TBLB

POSITION	CONNECT TO
1	vacant
2	RLY 2 , I and P
3	RLY 2, L and M

SH

POSITION	CONNECT TO
1	RLY 1, L and M
2	ammeter
3	ammeter

General wire connections

TBLA

POSITION	CONNECT TO	WIRE GAUGE, STRANDED
1	B1+	12
	SCR 1,2,3,4	18
	VR +	18
2	B1-	12
	AMAM 1 and 2	18
	VR-	18
3	Electrolyzer output +	12
	Relay 1	12
	VM+	18
4	Electrolyzer input -	12
	AVM	18

Process Controller

General wire connections

TBLB

POSITION	CONNECT TO	WIRE GAUGE, STRANDED
1	not connected	
2	pump out +	12
3	pump in + from B2+	12
	RLY 2	12

TBLC

POSITION	CONNECT TO	WIRE GAUGE, STRANDED
1	PRSW 1 and SCR 1 gate	18
2	PRSW 2 and SCR 2 gate	18
3	PRSW 3 and R2	18
4	PUSW 1B and SCR 3 gate	18
5	PUSW 2B and SCR 4 gate	18
6	PUSW 1A, PUSW 2A and R5	18

SH

POSITION	CONNECT TO
1	RLY 1, L & M
2	AAM
3	AAM
4	ESPM +

Wire assemblies

There are five ports in the process control unit. Three of these ports (ports 2, 3, and 4) are dedicated for wire assembly feed-throughs. Mark each wire in each assembly with its wire name.

Port 2, Wire assembly 1

WIRE NAME	WIRE GAUGE, STRANDED	CONNECTS TO	POSITION
1	18	TBLC	1
2	18	TBLC	2
3	18	TBLC	4
4	18	TBLC	5
5	12	TBLA	1
6	12	TBLA	2

Port 3, Wire assembly 2

WIRE NAME	WIRE GAUGE, STRANDED	CONNECTS TO	POSITION
1	18	TBLC	3
2	12	SH	
3	12	TBLB	3
4	12	TBLB	2

Port 4, Wire assembly 3

WIRE NAME	WIRE GAUGE, STRANDED	CONNECTS TO	POSITION
1	12	TBLA	3
2	12	Not connected	
3	12	TBLA	4
4	18	TBLC	6

Process controller box fabrication

The box we used for this project was just barely large enough for the circuitry. I recommend using a larger box, which would allow more room for wiring and make design and assembly a lot easier. Of particular concern was the wiring coming from the meters after assembly. These wires had a tendency to interfere with the operation of the relays' moving parts. This was addressed by shoving a few wires here and there, but it would have been better to not have the problem in the first place. If you have minimal experience fabricating such enclosures and do not like to work in tight spaces, then you definitely should use a larger box. Of course you could redesign and develop different routes for wires, connections, etc., for this size box, if you are so inclined.

Preparing the ports

The box has five ports (see illustration, next page). Ports 2, 3, and 4, are wire outlets. Port 1 is a pressure port which permits the box to be pressurized and depressurized by turning a ball valve on or off. We chose a low cost valve, but recommend using a better valve.

The case walls are thin plastic, so do not stress them too much when tightening fittings.

- Drill and tap four ports in the back of the box or elsewhere for ¼" NPT fittings
- Insert the three brass hose fittings, barb x male pipe for ½" hose ID and ¼" pipe into ports 2, 3, and 4
- Thread seal and tape the ⅛" hose x ¼" pipe male end of ¼" NPT male x ¼" NPT female with attached barb and male pipe fitting. Screw this into the female ¼" miniature chrome plated brass ball valve before screwing the ball valve into the case.

Practical Hydrogen Systems

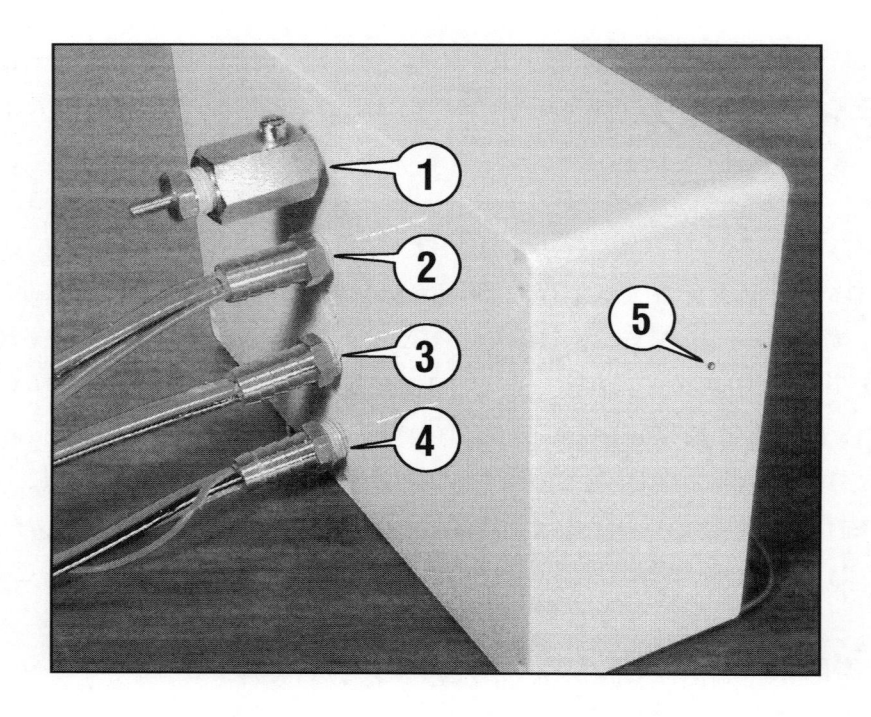

Location of ports

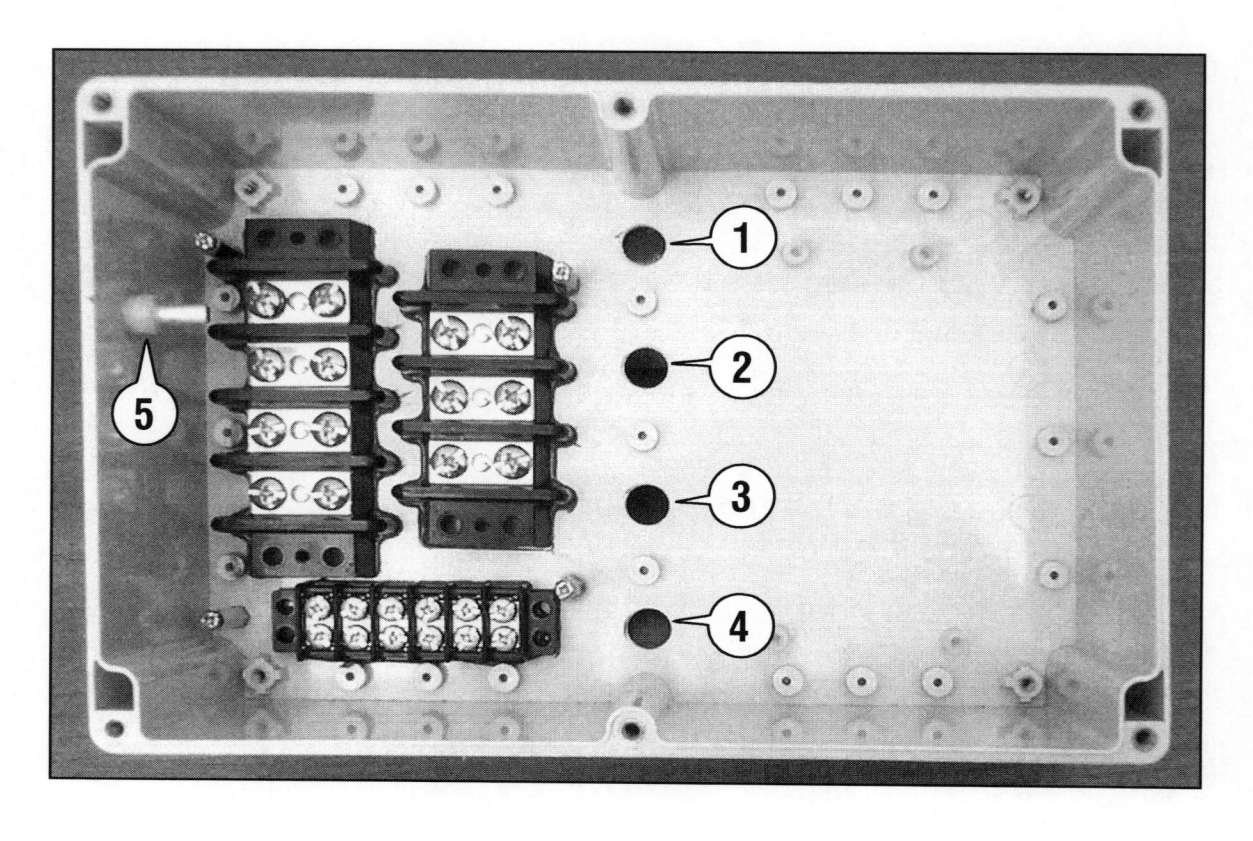

Process Controller

- Insert the ball valve, screw driver handle, ¼" NPT male x ¼" NPT female with attached barb and male pipe fitting for ⅛" hose ID, ¼" pipe male into port 1.
- Thread seal and Teflon tape the ¼" male end, and screw and tighten gently onto the box.

You can also epoxy the interface between the plastic box surface and the fittings on both the outside and inside of the box to help further ensure a leak free fitting.

Silicone could be used instead of epoxy here since it maintains flexibility better when exposed to temperature changes and weathering. Epoxy tends to delaminate when temperatures change and will expand and contract the connections.

Pressure sensor

There is one more pressure port – the inlet for the pressure sensor for the box. This consists of a small drilled hole in the box. Ours is located on the side of the box (port 5), but it can be anywhere convenient for attaching the pressure sensor hose.

Process control box pressure sensor port (above),

and with pressure sensor installed (below)

The pressure sensor is connected to its port via a plastic tube. Two standoffs are used to create a port attachment for the tube coming from the pressure sensor.

We used an aluminum standoff that slips tightly into the tube, and another plastic standoff that the aluminum standoff can be placed into. The wider plastic standoff provides firm support and gluing surface for attachment to the side of the box. The hole in the center of this plastic standoff was just the right size to accommodate the aluminum standoff.

Installing the pressure sensor

- ◆ Coat the outer part of the aluminum standoff that will be in contact with the plastic standoff with epoxy, insert into the plastic standoff, and let dry.
- ◆ When the epoxy has dried, align the plastic standoff with the port hole and epoxy into place, and let dry.
- ◆ When this has dried, insert a small plastic tube over the nipple of the sensor, and slip the other side of the tube onto the aluminum standoff.
- ◆ Coat the edges of the tubing with epoxy or silicone to seal.

If you can't find standoffs that fit each other for this purpose, either widen the hole in the plastic standoff or drill a hole in a small piece of plastic rod to accommodate the aluminum standoff. Or, you may come up with something else.

Wire ports

At some point in construction, draw the wires into the box through the brass ports, then seal these ports with epoxy or silicone. Flow the epoxy or silicone into the brass fittings around the wires so that no gaps allow air to enter or leave the system. This can be a bit tricky and may take several times to get it tightly sealed.

This is just one design for pressurized ports. You may well be able to come up with a more elegant approach.

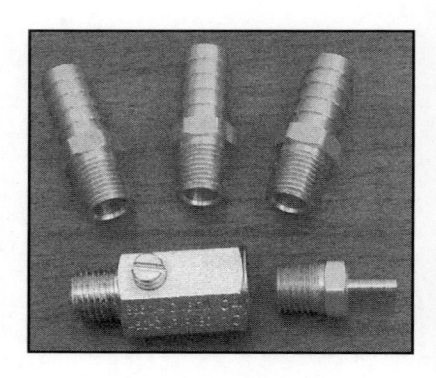

Pressure fill/relief port and wire port fittings

Process Controller

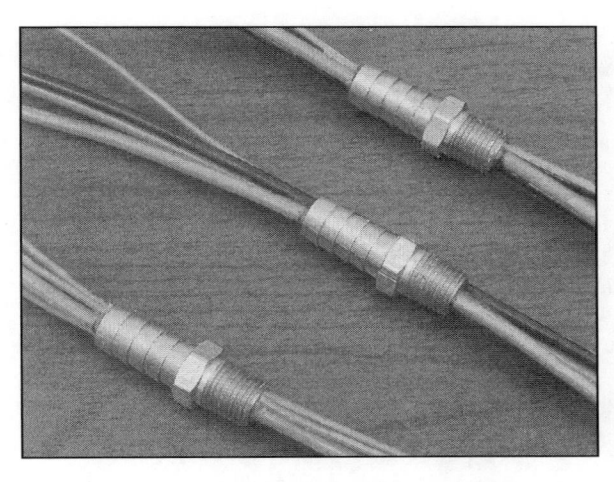

Wire assemblies with port fittings

Pressurize the box

- Attach a hand pump with a plastic hose to the ⅛" barb
- Open the valve and bring pressure up to about 2 or 3 psi.
- When the pressure target is achieved, close the valve.

If the box does not maintain pressure, try sealing the edges of the box with clear tape that covers the edges where they connect.

If testing shows that there is still a problem, ascertain where the leak is coming from and apply more epoxy or silicone. If you are using silicone and need more viscosity, you can cut it with xylene before pouring so that it will be more amenable to filling in all the small spaces.

Many systems have a small dedicated compressor to keep the enclosure's pressure up as a safety measure. Simple circuitry could be added for this. If the pressure falls to a preset level, a pressure switch will turn on the compressor to keep it above the set level.

Other details of circuit fabrication are shown in the photos. The physical circuit layout can vary according to the size and shape of the box used. We used several circuit platforms due to limited box space. The whole of the circuit could be laid out on one flat surface

Practical Hydrogen Systems

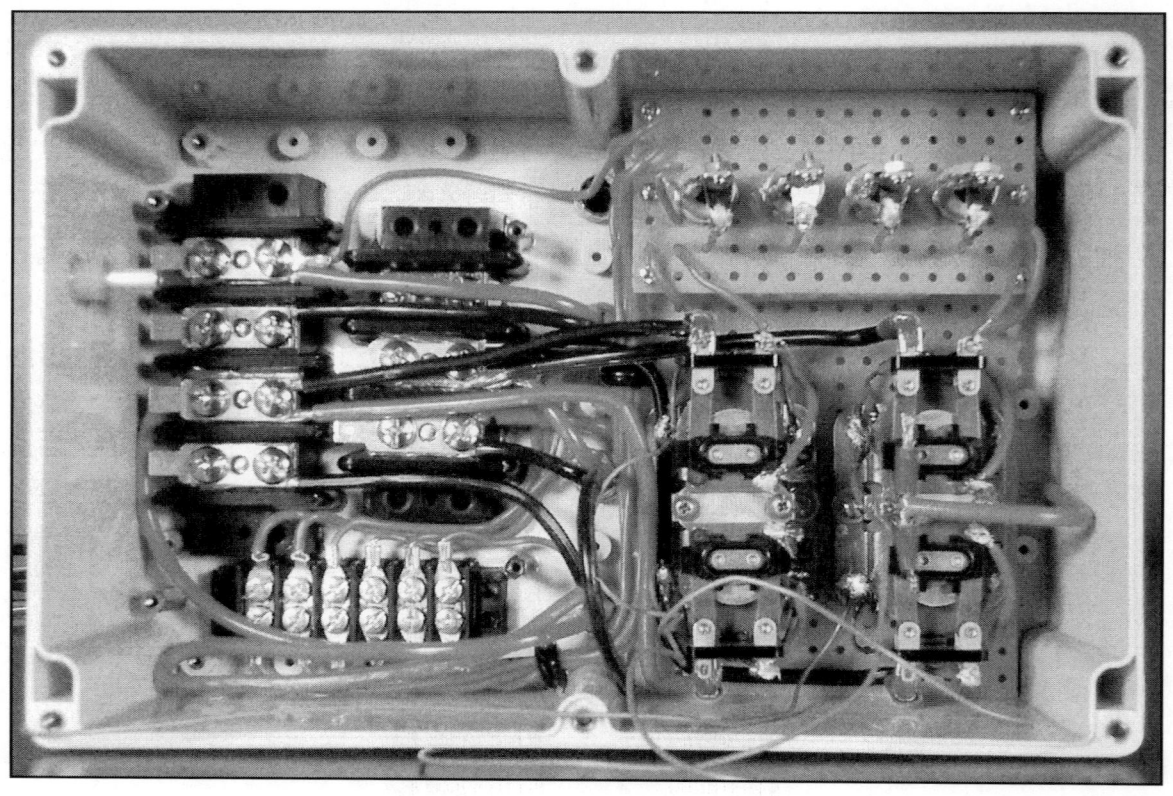

Process Controller

with enough space to make soldering and making connections comfortable and easy.

Test all circuits before placing them in the box to get a good idea of how each particular circuit works.

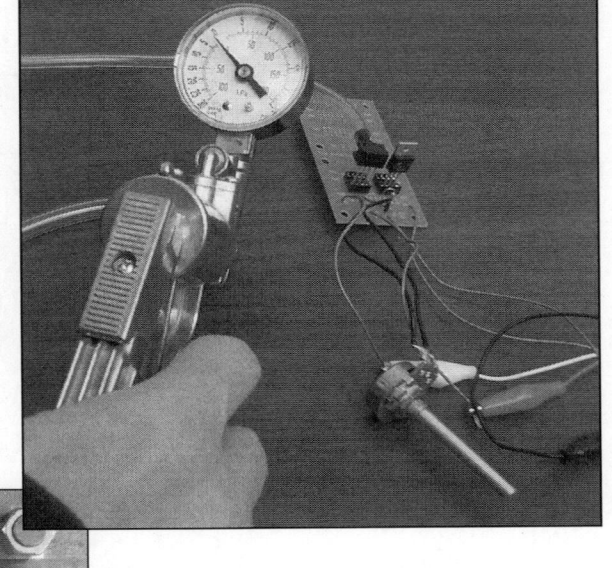

Most of this circuit could have been replaced with a micro-controller, which would have reduced needed space dramatically. We chose to use this circuit initially and wait for the next project to try out such improvements.

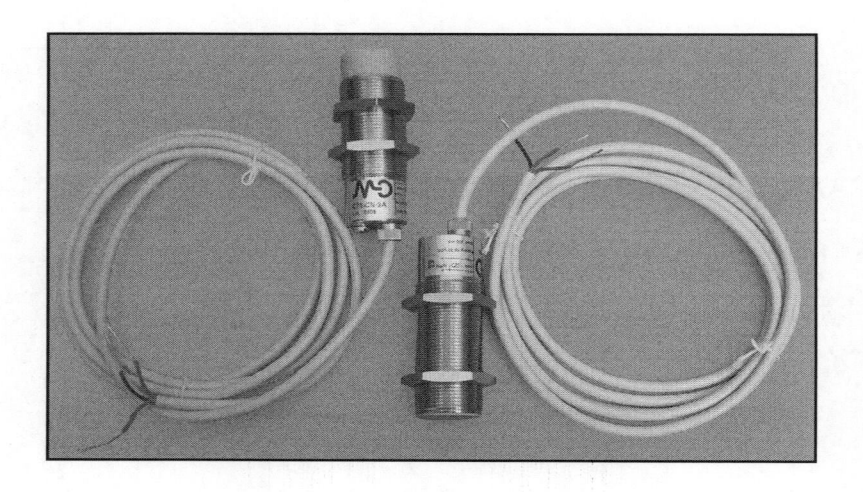

Capacitive sensors

Capacitive sensors are one option for detecting the electrolyte level in the system. They have the ability to detect nonmetallic materials such as electrolyte solution and are available in either normally-open (NO) or normally-closed (NC) circuit configuration.

Normally open means that when the unit is powered up it will not pass current through the trigger wire until the target material is detected. The normally-closed circuit configuration allows current to pass through the trigger wire until the target material is removed.

The detection range of the particular capacitive sensors that we used is about 0 to 20 mm. They have a sensitivity adjustment with red and green LED indicators. When the normally-open unit is powered up at around 12 volts, the green LED lights up indicating a "power on" state. The red LED remains off. When the target material is detected, the red LED is activated and both the red and green LED are lit.

In the normally-closed configuration, the green and red LED remain lit when there is no target material present. When a target material is present the red LED goes off and only the green remains on.

Process Controller

Sensor configuration

Although capacitive sensors can be wired in a number of ways to accomplish the same end, we decided to use one normally-open and one normally-closed sensor to turn the pump on and off.

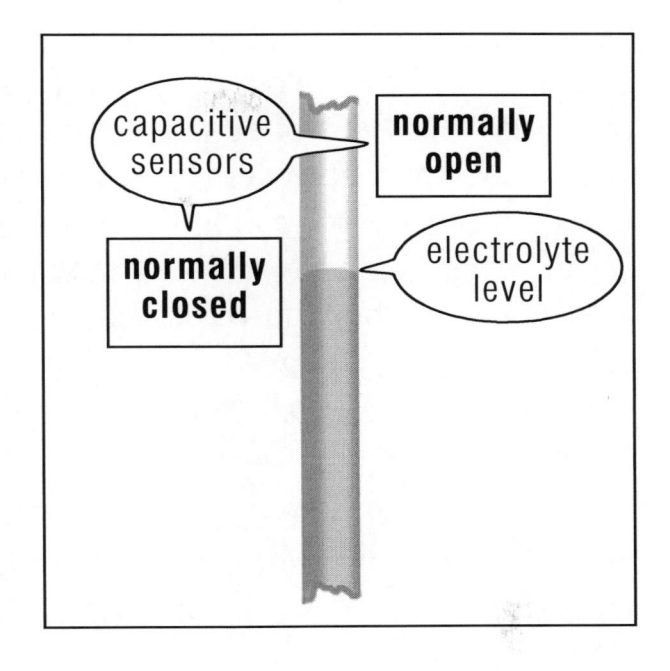

The normally-closed sensor was used for the low level indicator which signals the pump to turn on. The normally-open sensor was used as the high level indicator and signals the pump to turn off.

When the water level is at the full position neither sensor activates. When the water/KOH level falls below the detection range of the normally closed sensor the pump is turned on and begins to fill the tank and electrolyzer. When the water/KOH line reaches the detection range of the normally open sensor the pump is turned off. This arrangement, as shown in the illustration above, works quite well.

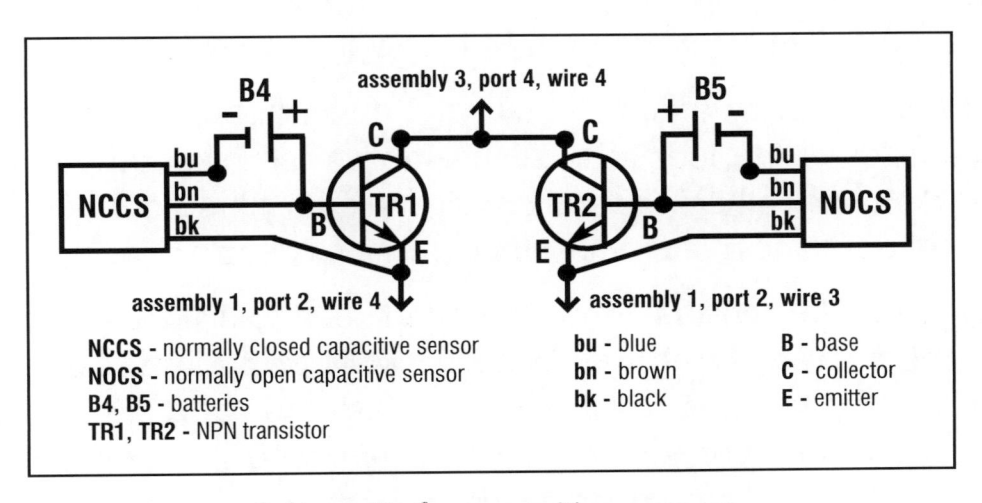

NCCS - normally closed capacitive sensor
NOCS - normally open capacitive sensor
B4, B5 - batteries
TR1, TR2 - NPN transistor

bu - blue
bn - brown
bk - black

B - base
C - collector
E - emitter

Schematic for capacitive sensors

Each sensor is coupled with an NPN transistor that acts as a switch that triggers the SCR and relay in the process controller. This type of sensor operates on 10 to 30 volts, however it is advisable to regulate the voltage input to the sensor at 10 volts for this particular system. The reason is that the maximum gate voltage on the SCRs used is 10 volts, so we used an LM 317 voltage regulator circuit to power the sensors at 10 volts.

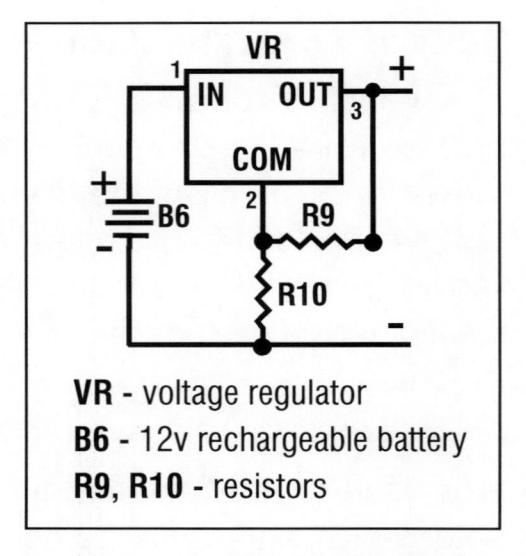

VR - voltage regulator
B6 - 12v rechargeable battery
R9, R10 - resistors

Optional voltage regulator for capacitive sensors

A straight 12 volt battery supply could be used to power the sensors. This should work well with the SCR since there is a voltage drop in the sensor, however we did not get around to testing this. In fact, during initial tests we used a fresh 9 volt battery which delivered about 9.5 volts, to power each sensor, and this worked fine.

Although this sensor fits into the system, it delivers more current and voltage than our ideal, which is to use 20 milliamps and around one or two volts in the external circuit. We wanted to try these sensors, to see if they would work, and if it would be worth designing a circuit to better integrate them into the system at a later date.

The sensors can be mounted any way you wish, but they must be on opposite sides of the tube, one slightly above the other, and must not interfere with each other in operation. Fill and empty the sight glass to test for optimal positioning and adjust their positions accordingly.

These particular sensors have a threaded body so distance adjustment can be made by screwing the sensor forward or backward on a base plate that is the thickness of the threads. They are not rated intrinsically safe or explosion-proof, so we cannot recommend them for use as is. They would have to be embedded in a pressurized and/or explosion-proof enclosure of some sort to meet this qualification. This would be fairly easy to fabricate.

Fiber-optic sensors

The best sensor option is an intrinsically safe fiber-optic photonic sensor system, either infrared or visible spectrum.

Light-emitting diodes or a laser are used in a fiber-optic conduit in a reflective or through-beam mode to signal the liquid level in the sight glass. The light is frequency modulated to reduce triggering of the sensors by ambient light or infrared radiation. Liquid level detection can occur with the presence of, or lack of presence of, the different reflective qualities of the liquid itself; or with the aid of an opaque float to trigger the signal by either interrupting or allowing the beam to pass through the sight glass. The fiber-optic conduit makes it possible for all active electrical components to be housed at a distance from the hydrogen system.

Other types of sensors

Hall sensors can be triggered by a ferrous float in the sight glass. We have not tried this technique so cannot vouch for its viability. You will have to experiment and see if this works.

At an early stage of the project we tried using a reed switch/magnetic float device that looked very promising on paper, but when we tried the technique out, the laws of physics got in the way. We quickly found out how tenaciously magnets can cling to reed switches.

Another possibility is acoustic/ultrasonic sensors. We have not experimented with these, so we do not have any information to offer in this regard.

Most of these different types of sensors can be internally and/or externally positioned. Internal positioning of sensors requires more fabrication work and a customizing of the sight glass for probes and attendant components

Practical Hydrogen Systems

In general, level switches should be repeatable, reliable, and of a decent life cycle. The ambient conditions that the switch will have to operate in must be considered. Capacitive proximity sensors, for instance, are affected by moisture. Infrared sensors are subject to false triggering by ambient sunlight, and so on. There are remedies for these situations, but it is good to know the characteristics of each particular type of sensor, and their good and bad points in order to design the most trouble free system possible.

A good basic introduction to the construction of a wide variety of sensors is "Electronic Sensor Circuits and Projects" by Forrest M. Mims.

Sensors materials list

Description	Supplier	Part #	Quantity
NOCS - 30 mm diameter, 10-30 volt DC, 3 wire, NPN, normally open, shielded, adjustable, capacitive proximity sensor	Automation Direct	CT1-AN-1A	1
NCCS - 30 mm diameter, 10-30 volt DC, 3 wire, NPN, normally closed, unshielded, adjustable, capacitive proximity sensor	Automation Direct	CT1-CN-2A	1
TR1, TR2 - 2N2222 NPN switching transistor	Electronics supplier		2
B4, B5 - Rechargeable 9V battery or regular alkaline	Electronics supplier, hardware store		2
(Optional)12V rechargeable to be used with optional voltage regulators			2

Process Controller

Sensors materials list

Description	Supplier	Part #	Quantity
VR - LM 317T voltage regulator	Jameco Electronics, electronics supplier	23579CE	1
R9 - 220 ohm, ½ watt resistor; R10 - 1600 ohm, ½ watt resistor; Resistor values for voltage regulator w/ 10.34V output	Ocean State Electronics	R9, R10	
(Optional) R9 - 220 ohm, ½W resistor; R10 - 1500 ohm, ½W resistor; Resistor values for 9.77V output	Ocean State Electronics	R9, R10	

Parts numbers and suppliers have been provided for your convenience; however, suppliers may go out of business and parts numbers may change. All parts listed are available from multiple suppliers.

Part 3
Hydrogen System Operation

System Setup

Frame

Setting up the system requires some type of frame. We used a hydraulic press frame and modified it to suit our system configurations as needed.

Whatever type of frame you use, the frame must be level, and the system must be level within the frame. The sight glass is calibrated to indicate the level of the KOH in the electrolyzer. If the system components are not level, and not in the same relative positions as when the sight glass was calibrated, the sight glass will not give an accurate electrolyte level reading.

Electrolyzer fluid level

It is very important to keep the fluids in the electrolyzer at the fill level. The fill level is about ¼" to ½" from the inside top of the flange that the duct is attached to. The electrolyte fluid must be well above the bottom edge of the PVC hydrogen duct coupler at all times so that the hydrogen

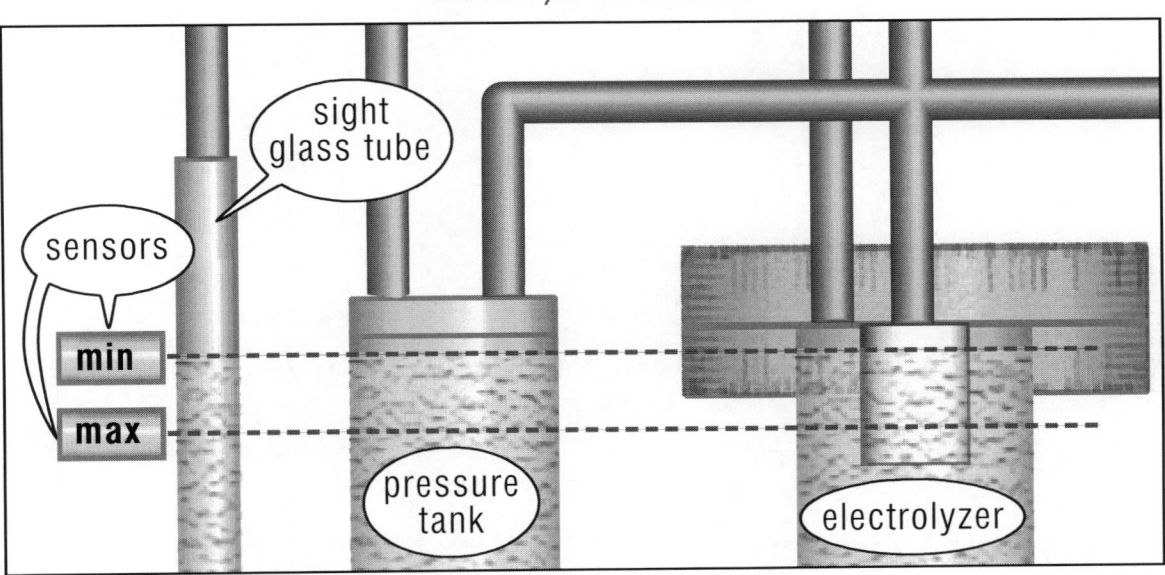

Electrolyte fill levels

and oxygen gases do not mix. The electrolyte reservoir must always be kept filled with enough liquid to feed the unit.

The pressure tank must be situated so that the fluid level is also ¼" to ½" from the top of the inside of the pressure tank. Be sure that the pressure tank and electrolyzer are situated correctly in relation to each other.

To calibrate the sight glass, find and mark the fill point on the sight glass. It should be level with a point that is ¼" to ½" below the bottom of the top flange in the electrolyzer, which should also be level with a point ¼" to ½" below the inside of the top of the pressure tank. This is easy to do and is vitally necessary to the safe operation of this electrolyzer.

As the water is used up in the system, the level will drop and the automatic sensor system will refill the pressure tank and electrolyzer, or you can activate the refill manually. When the sensors are set up for this system, place them closely so that the level never falls near or below the edge of the hydrogen duct. As you experiment with the system you will find the proper spacing and positions for the sensors, so that the system can operate safely.

Operating Procedure

Establish the routine

Once the components of the hydrogen generating and processing system are constructed and assembled, make a list of the routines necessary to start and operate the unit.

An operating check list is an absolute necessity for any hydrogen installation. There will be times when your attention is elsewhere, or focused on the outcome of a specific experiment. Put quite simply, people make mistakes. Problems can be avoided by routinely running through the check list at critical points, such as initializing the system, routine online and off-line procedures and emergency shutdown procedures.

Although your routine will vary according to your design and construction, the list below is our check list for the system in this book.

Practical Hydrogen Systems

Initial startup

1. Put on protective gear
2. Check all fittings to see if OK
3. Check to see that unit is level
4. Check wiring to see if everything is connected appropriately
5. Manual switch to panel off
6. Manual switches to sensors off
7. Open oxygen and hydrogen valve. Before opening hydrogen or oxygen valve, always attach a long tube to the outlet. If there is some pressure in the system the gas will expel and then may pull fluid KOH out. With the tube, any fluids will be directed to where you want them to go. To attach plastic tubing to Yor-Lok fittings, simply slip tube into fitting and tighten slightly.
8. Fill bubbler to correct level
9. Relieve water in coalescer if necessary
10. Relieve tank of water if necessary
11. Check reservoir to see if it has fluid
12. Open pump valve
13. Manually start pump to fill KOH to a fill level mark
14. Turn pump off
15. Close pump valve
16. Close oxygen valve
17. Purge (see nitrogen purge section)
18. Close hydrogen valve
19. Check to see if the fluid level is the same or has changed. If your fluid level is below the fill point mark, open the pump valve and turn on the pump momentarily with the jog switch until the fluid level reaches the point it needs to be at. If the fluid has exceeded the

Operating Procedure

fluid level, open the three way valve between the pressure tank and the pump to discharge some fluid. Be sure to have a hose placed on the outlet of the three way valve so that when you discharge any fluids under pressure they are aimed well away from anything so that the KOH is discharged in a safe manner.

20. Open valve to process controller
21. Connect magnetic switch in process control box
22. Pump up process control box to 2 psi
23. Close valve to process controller
24. Turn on panel switch
25. Turn on sensor switches
26. Check readings in process controller
27. After a short period of operation, check sight glass fill mark level and adjust if necessary by jogging pump or releasing KOH through three way valve

In process

1. Observe for correct operation

- Observe oxygen gauges and hydrogen gauges. Hydrogen should be at the same pressure or have slightly higher pressure than oxygen. If not turn off system and adjust, then restart.
- Observe pressure readings
- Observe electrical readings
- Visual check for gross leaks
- Gas detector check, note sounds etc.
- Check sight glass fill mark level

2. Tank filled

- Depending on how you construct the system, either leave the system in automatic mode or physically attend to certain functions. If you are using an equalization valve, simply use the gas directly from the tank without going through the following adjustments. If you are using the accumulator the routine might be similar to this:
- Turn off panel power manually
- Turn off sensor power
- Release oxygen to zero psi. Hydrogen will cross over and release.
- At this point, either open the hydrogen valve; or as you are releasing the oxygen, leave about 1 psi in the hydrogen side and then close oxygen valve
- You can now use gas directly from the storage tank, or transfer the gas to another tank

You have several options at this point depending on what you have done.

- If you use the gas from the storage tank directly and have exhausted the gas you can then drain the tank, or wait for a few more cycles of fill and deplete before you drain.
- If you do not drain the tank and leave about 1 psi on the hydrogen side, then check KOH level, pump more in if needed, and then close the oxygen and hydrogen valves. Manually reconnect the power source and sensor switches for another run.
- If you released the hydrogen (opened the hydrogen valve) then purge the tank and the hydrogen side of the system before restart. (2 to 5 psi)

After you have worked with a system a bit, you will better understand what your particular needs are, and you can develop your own routines.

Calculations for Production and Storage

The gas output of the electrolyzer can be tested either with field tests or bench tests. Field and bench tests using a constant source of DC for the electrolyzer will give a fairly accurate picture of electrolyzer performance and efficiency. Field and bench tests that do not use a constant source of DC but rely on intermittent power supplies such as photovoltaic panels and wind turbines, etc., will give variable inputs depending on the intensity of the power source (for example, sunlight or wind), available.

When using intermittent power sources, data log variations in input to the electrolyzer and average these figures to calculate the power input.

Measuring gas output

To calculate the general output of hydrogen and oxygen from your electrolyzer, measure the voltage at the electrolyzer terminals during operation. Measure the amperage at the same time by breaking the circuit on the positive side with one lead of your meter or data logging apparatus going to your power source, and the other lead on the positive terminal wire for the electrolyzer. Multiply the voltage times the amperage to get the power draw of the electrolyzer in watts. For instance, if the electrolyzer draws 2 volts at 20 amps, this would be 40 watts of power being used by the electrolyzer.

Next figure out how many joules are being used per second. Basically 1 watt per second equals 1 joule per second. 40 watts of power usage would be equivalent to 40 joules per second. There are 3600 seconds in an hour, so multiply 40 times 3600 seconds. This equals 144000 joules per hour (144 kj).

One liter of water yields 1,358.3 liters of hydrogen and 679.15 liters of oxygen. It takes 13,170.9 kj to disassociate one liter of water. So we divide 13,170.9 kj by 144 kj which indicates that it would take 91.46 hours to electrolyze 1 liter of water and produce 1,358.3 liters of hydrogen and 679.15 liters of oxygen.

To find out how many liters per hour of hydrogen will be produced, divide 1,358.3 liters by 91.46 hours. The result is the liters per hour that will be produced, and in this case would be 14.85 liters of hydrogen per hour. Similarly, to find out how many liters per hour of oxygen will be produced, divide 679.15 liters by 91.46 hours. The result is the liters per hour that will be produced, in this case, 7.42 liters of oxygen per hour.

To work with cubic feet of gas rather than liters, divide by 28.317. For instance, 14.85 liters of hydrogen divided by 28.317 equals 0.524 cubic foot per hour or about one half cubic foot per hour. And, 7.42 liters of oxygen divided by 28.317 equals 0.26 cubic foot per hour.

Oxygen production will always be one half of the amount of hydrogen production. These volumes are calculated for what is termed SLC (standard laboratory conditions) which is considered to be 24.47 liters at 25°C or 298°K, and at a pressure which is one atmosphere or 101.3kPa. You can also calculate for what is called STP (standard temperature and pressure). This is considered to be 22.4 liters at 0°C (273°K) and 101.3kPa (one atmosphere).

More precise results require using the actual ambient temperatures involved during the data collection process as well as the integration of other factors into the equation, however this will suffice for general ballpark purposes.

General voltage efficiency

To calculate the general voltage efficiency of the electrolyzer, note the voltage draw and divide 1.24 by this voltage draw. So, if the measured voltage at the electrolyzer terminals is 1.8 volts, divide 1.24 volts by 1.8 volts. This will give you a figure of 69% which is the general voltage efficiency.

Calculations for Production and Storage

Please note that this does not tell the whole story and is not the complete efficiency figure for the electrolyzer.

Calculating tank capacity

To calculate the cubic foot capacity of a cylindrical tank, multiply 0.79 times the diameter. Multiply the result by the diameter again, then multiply that result by the length of the cylinder. For instance, for a drum or container 7 feet in diameter and 10.4 feet in length, the calculation and result would be: 0.79 x 7 x 7 x 10.4 = 402.584 cubic feet. To convert this to US gallons, multiply 7.5 x 402.584 (there are roughly 7.5 US gallons per cubic foot). So the total gallon capacity for this tank is about 3019.38 US gallons. The above equation gives a ballpark figure for storage at atmospheric pressure (14.75 psi). If you are considering a higher pressure storage system, such as around 60 psi, you could store about four times as much gas in the above example in the same size tank.

Working with Commercial Hydrogen Cylinders

Hydrogen gas can be delivered in a variety of ways, from bulk tankers and tube trailers to cylinders and lecture bottles.

For most purposes, a 31" or 51" size cylinder is sufficient for most applications. The 31" cylinder contains about 65 cubic feet of gas and weighs about 65 pounds. The 51" cylinder contains about 196 cubic feet of gas and weighs about 133 pounds.

Hydrogen can also be stored in small lecture bottles, which are small steel cylinders about 2" in diameter and 15" long. They use a CGA 170 or CGA 180 regulator and valve connection. These are used mainly for demonstration purposes where a small portable supply is needed. They can be easily recharged from larger cylinders.

Cylinder pressure is around 2200 psig at 70°F. They are equipped with a CGA 350 valve.

Hydrogen purity grades

Hydrogen can be purchased in various grades of purity. Research grade is 99.9995%, ultra high purity is 99.999%, high purity is 99.99%, and industrial grade is 99.95%. Of course, the purer the gas, the more expensive it is.

Most applications, such as fuel cells, will do well with industrial grade hydrogen. Many commercial units operate on this grade, although some manufacturers may require a more pure gas.

Purchase or rent the gas cylinder

Gas is supplied and delivered by a variety of companies that cater mainly to the welding, medical and industrial trades. A yearly rental fee is charged for the cylinder, and for a small fee you can bring it in and/or have them pick it up and recharge the cylinder for you. In some areas a lead time of about a week is needed as hydrogen may not be used often in some locals. For instance, in the rural area where I live, there is almost no industrial use of hydrogen, so the local supplier has to request a delivery. When the delivery arrives I simply bring the discharged cylinder in and pick up a fully charged one.

You can purchase your own cylinder with a CGA 350 valve, however, for safety reasons, you will be responsible for having it retested every five years, or according to current regulations.

The smaller 31" cylinders are easier to handle. Most people prefer these if their gas consumption does not quickly reduce the contents of the cylinder. We prefer to have several smaller cylinders for experimental purposes, largely due to the ease of handling.

All cylinders come equipped with a valve specific to the gas in the cylinder, and a cylinder cap. The cylinder cap prevents damage to the cylinder valve while it is in transport, or when it is not in service. If a cylinder were to be tipped over and dropped, damage to the valve at worst could create a dangerous projectile, and a serious fire and explosion hazard. All cylinders in service should be secured by straps or chains, and set on a untippable base, or holding cage, or gas cabinet so that they can not fall or be knocked over.

Storing and moving hydrogen cylinders

Hydrogen cylinders should not be stored near oxidizers, and other flammable substances. They should be stored in a cool dry place that is well ventilated, fire resistant and in compliance with federal, state and local regulations. Cylinders should always be stored in an upright position, never on their side, and they should not be rolled on the ground when moving them. The best and easiest way to move a cylinder is with a cylinder cart.

Working with Commercial Hydrogen Cylinders

Never reduce the cylinder pressure below the operating pressure of your system. Usually it is recommended not to go below 25 psig, or at the very minimum 7 psig. You can use check valves to prevent reverse flow into a cylinder.

Although not necessary, cylinder wrenches can come in handy as some caps may have been placed very tightly on the cylinder making it difficult to unscrew. The cylinder wrench will make easy work of the matter. These wrenches will also help to open very tight cylinder valves easily.

Removing the cylinder cap

Hydrogen regulators, cylinder valves and outlet connections

Regulators are used to reduce cylinder supply pressure to a lower use pressure. As stated previously, hydrogen uses a CGA 350 cylinder valve. This requires a CGA 350 outlet connection on the regulator. Be sure that the regulator you purchase for cylinder use has a CGA 350 outlet connection. In line hydrogen regulators can have any variety of connections, but cylinder regulators must have the CGA 350 inlet connector.

There are basically two types of regulators, single-stage and two-stage. Single stage regulators control pressure output of gases, but delivery pressure increases as the cylinder pressure decreases, which requires constant readjustment of pressure control to maintain constant delivery pressure.

Practical Hydrogen Systems

Cylinder regulator

A two-stage regulator moderates the flow of gases in the second stage, which keeps the pressure constant as the cylinder pressure decreases. This eliminates the need for constant pressure control adjustment. If it is critical to maintain a consistent pressure from the cylinder, for instance, as a supply to a fuel cell unit, then you will need a two-stage regulator. Two-stage regulators cost more than single-stage regulator.

New or used regulators

Regulators can either be purchased new or used. Any used regulator should be well inspected before service. A lot of older regulators sometimes look brand new, and indeed may not have been used very much or at all. However, what is generally referred to as the rubber goods or soft goods (seals) may have deteriorated over time, which would make the regulator a hazard. There is no way to test this until you get the regulator in hand.

Some people will tell you that they have used the regulator and it worked fine for them. This is no assurance of a faulty or leaky regulator. If you can, purchase a new regulator for the critical applications in your system and for cylinder use. This does not mean that you cannot find a perfectly serviceable used regulator at a more pleasing price, just be aware that the equipment could be faulty and unusable; or, worse, faulty and dangerous.

When purchasing cylinder regulators make sure that the high pressure gauge reads higher than the cylinder pressure of the gas. Most gauges go to 3000 or 4000 psig. This is fine for hydrogen cylinders which are usually pressurized to 2200 psig.

Working with Commercial Hydrogen Cylinders

The outlet gauge should have gradations fine enough for your purposes. There is nothing more annoying than roughly estimating between tick marks on the gauge if exacting outlet pressure control is needed. If you need a supply pressure of 2 to 6 psi you will not want a gauge that reads to 500 psi. You will want a gauge that has the smallest span possible where the gradations are clearly and visibly marked for your particular needs.

Generally speaking a 0-100 psi gauge will meet a wide range of needs, unless you are refilling hydride bottles which will require higher pressures. A 0-500 range would be more suitable in that case. If you have a number of different applications where the outlet pressure used is quite different, you will want to purchase a regulator for each specific operation.

Besides hydrogen cylinder use, single- and two-stage regulators are available as line regulators, and high-purity regulators. Line regulators are placed at any point in a system where they are needed. High-purity regulators minimize out-gassing and diffusion. Basically for hydrogen systems you will be using general purpose single-stage line regulators and two-stage cylinder regulators.

Most regulators are equipped with a relief valve, one or two outlet ports and an inlet port. The inlet port of a hydrogen cylinder regulator has to have a CGA 350 valve connector which consists of a bullet shaped nipple and a nut. The outlet fitting on the regulator can vary and can be changed and/or adaptors used according to need. General-use line regulators can have a variety of outlet and inlet fittings and can be changed or adaptors used to fit requirements.

On the outlet side of the regulator, you can place a check valve to provide back pressure protection. This will also help to prevent damage to the regulator if high pressure is used in your system. At low pressures regulator damage is not so much of a concern, but you may want to prevent back flow if you inadvertently reduce the pressure in your cylinder below the pressure in the system. Also, for further output control, you can place an on-off valve near the outlet. This is an added safety measure, and always a good idea.

Practical Hydrogen Systems

Connect cylinder and regulator

To put your cylinder into service:

Connect regulator to cylinder, then tighten connection between regulator and cylinder

1. Connect the regulator to the cylinder.

2. Seat the nipple into the valve on the cylinder and screw the nut, hand tight.

3. Apply pressure with a wrench to secure tightly. As you tighten the nut, it will seal the surface of the nipple with the valve outlet where they touch. Sealing does not occur along the

Working with Commercial Hydrogen Cylinders

threads with this type of connection. Recommended torque for a CGA 350 is about 35 ft-lb with a maximum of 50 ft-lb.

4. With the regulator secured to the cylinder, make sure the outlet valve is closed by turning the adjusting handle to the full counter-clockwise position on the regulator. The regulator out-

Close regulator outlet valve, then open the cylinder valve

let valve must always be closed before the next step, which is opening the cylinder valve.

5. To open the cylinder valve, place both hands on the cylinder wheel and turn on slowly, allowing the pressure to gradually rise in the regulator. When the high pressure gauge shows maximum pressure (about 2200 psig), open the cylinder valve wheel fully.

Practical Hydrogen Systems

6. To adjust the outlet pressure, slowly turn the adjusting handle on the regulator in a clockwise direction until the outlet pressure gauge indicates the pressure you desire.

7. When you have finished using the gas, always turn the cylinder valve wheel on the cylinder to the off position.

Adjust outlet pressure

8. When the cylinder valve is closed, then disconnect or isolate your low pressure equipment from the regulator

9. Vent the gas from the regulator by turning the pressure adjustment handle on the regulator to a clockwise position, thus discharging the gas.

It is important to isolate the regulator and or disconnect the regulator feed hose from your system equipment, because when you turn the adjusting handle, the pressure will rise above

Close cylinder valve

256

Working with Commercial Hydrogen Cylinders

your intended delivery pressure. If your equipment is pressure critical, this can damage it. For instance, some fuel cell units take a pressure of from 4 to 6 psig. When you vent the regulator the pressure will rise to perhaps 10 or more psig and it can blow out the membranes of the fuel cells from over pressure, or create other problems.

Nitrogen Purging

Working with commercial nitrogen cylinders

For purging operations, use nitrogen cylinders. These have a CGA 580 valve and fit a CGA 580 nipple and nut connector on the regulator.

Some suppliers prefer to sell small nitrogen cylinders and then exchange cylinders when you need a refill, rather than renting the cylinder. This will save you money on cylinder rental fees. A small 40 cubic foot cylinder usually suits our needs.

Recommended torque for a CGA 580 is 40 foot pounds with the maximum being 60 foot pounds.

Nitrogen is an inert gas and does not readily react with other materials, so it's compatible with most materials. It is not flammable or toxic, however it can be an asphyxiation hazard if it is used in an enclosed space and it replaces the air that you are breathing. It is also a hazard because it is a gas in a cylinder under high pressure.

Nitrogen tank for purging

Purging apparatus

Purging apparatus for this particular system was kept simple, though there are many ways to purge a system of contaminants. A permanent purge fixture can be attached to the system, or simply use various ports on the system as purge inlets and outlets.

Two types of permanent assemblies are used for purging. One is the T type and the other is the cross type. Purge assembly components can either be purchased or made from available valves and fittings.

Purging techniques

The best purge method is to evacuate the system and then fill the system with an inert gas such as nitrogen.

Second best is to run nitrogen through the system until all contaminants are expelled and replaced with nitrogen.

A third method is to run the system, and allow the created gas to purge the system. This is not recommended and is used mainly to purge short lengths of tubing for filling hydride bottles. It is dangerous because it can create an explosive concentration of hydrogen and air. If a system has a catalytic recombiner, do not purge with generated gases. Remember that any platinum catalyst will ignite a hydrogen and air mixture, or hydrogen and oxygen mixture.

No matter what method is used to purge, the purge process should be repeated at least six times.

Purging can be done from 2 to 20 psi, though purging can be done at any pressure desired, as long as the system will allow and contain the pressure.

If the system is shut down and there is a total loss of pressure (it leaks), the system should be purged again before use. If the leak allows some pressure to remain in the system, it's not necessary to purge, but you should search and tighten. Any time there is a breach in the system where oxygen or air may have gotten into the hydrogen stream, you should purge.

Consider nitrogen purging a fundamental and necessary task for operating a hydrogen system.

Nitrogen Purging

Nitrogen purging sequence

1. Put regulator with attached outlet hose on cylinder and tighten CGA connection.

2. Place hose in hydrogen outlet on electrolyzer and tighten.

Practical Hydrogen Systems

3. Close oxygen outlet on electrolyzer (left).

4. Open hydrogen outlet on electrolyzer (right).

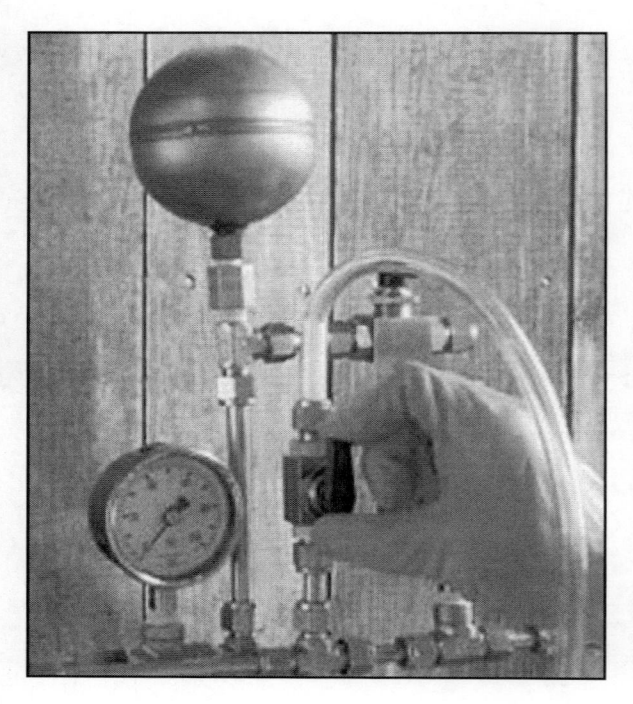

5. Open high pressure cylinder wheel (below left).
6. Slowly open low pressure outlet to 6-10 psi (below right).

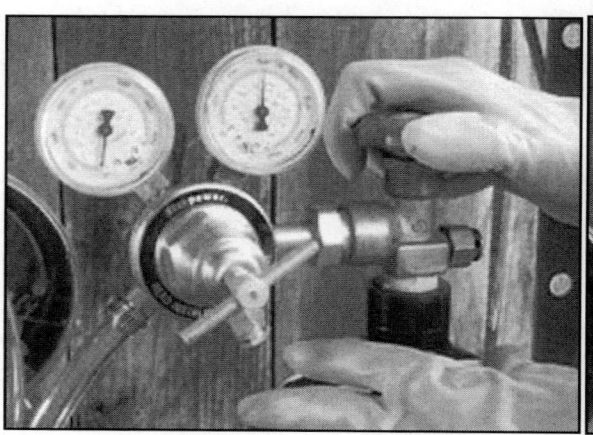

Nitrogen Purging

7. Purge gas from system and tank six to ten times by opening regulator valve on storage tank.

8. Close cylinder valve.
9. Close hydrogen outlet valve on electrolyzer.
10. Remove hose from hydrogen outlet on electrolyzer.
11. Vent gas that remains in regulator – low pressure clockwise.

The purge routine will generally be dictated by the configuration of the system. The idea is to remove all contaminants from the process gas. In the case of an electrolyzer system, this means the removal of oxygen impurities from the hydrogen stream, and hydrogen from the oxygen stream. Particular attention is paid to the hydrogen stream since this is the gas that will be stored and used. To store hydrogen safely, oxygen contamination should be at a minimum.

Hydrides for Hydrogen Storage

Hydrogen storage methods

The most common method for storing hydrogen is to apply high pressure, which reduces the space necessary for storage. Hydrogen can also be liquefied and stored in special dewar double-walled containers at a temperature of -253°C.

Other types of storage methods have been developed and continue to be developed. These include the use of metal organic frameworks, nano tubes, metal hydrides, glass micro-spheres, polymers, hydrogen clathrate hydrates, zeolites, and so on. All of these methods have some benefit and further development will surely bring a number of adequate storage technologies in the very near future. At present the emphasis is on metal hydrides for a practical near-term solution relative to the other storage methods under scrutiny.

Certain metals can absorb and store large quantities of hydrogen in the interstices of their crystal lattice structure at relatively low pressures. The compound created by this metallic absorbent and the absorbed hydrogen is called a hydride.

Pros and cons of hydride storage

The storage of hydrogen in hydride form is a safer and more compact alternative to the very low temperature storage requirements for liquid hydrogen, and the high pressure needs for gaseous storage. Hydrides have the advantage of a higher density storage than either pressurized gaseous hydrogen or liquid hydrogen.

On the down side, hydriding breaks down the metal crystals over a period of charging and de-charging cycles. The metal expands and contracts, which eventually turns the crystals into a fine powder, and reduces their storage capacity, to the point where the metal or alloy has to be replaced. Another disadvantage is that these metallic sponges prefer 99.95% pure gas. Both water and oxygen and other impurities in the gas stream will accelerate the deterioration of the absorption capacities of the hydride over time, so it is important to keep the gas stream as clean as possible. Using hydride storage requires more attention to removing contaminants from the generated hydrogen.

Types of hydrides

There are basically three types of hydrides: ionic, covalent and transitional metal-interstitial hydrides.

There are many alloy compositions available. The most common are alloy types:

- AB – iron and titanium
- AB5 – lanthanum and nickel
- A2B – magnesium and nickel
- AB2 – zirconium and vanadium

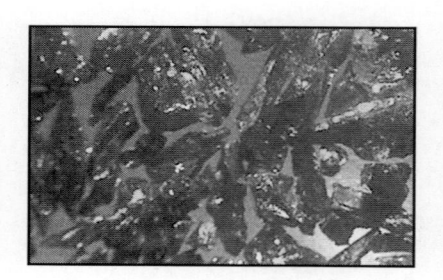

Metal alloy before hydriding (above), and after hydriding (below).

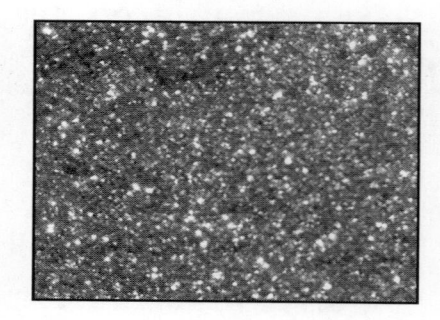

Each type has its own particular pressure and temperature characteristics which are suitable for different purposes.

The most common alloy used for hydrogen storage experiments is AB iron/titanium. This combination has the capacity to hold about 1.95% of its weight in hydrogen. Iron/titanium permits charging (adsorption) and de-charging (desorption) hydrogen at ambient temperatures and relatively low pressures.

Although other hydrides have greater storage capacities, iron/titanium types are a good introduction for working with hydrides for the experimenter.

Hydrides for Hydrogen Storage

Hydride bottles

A hydride bottle, complete with attendant valves that can be easily connected to whatever apparatus you want to feed hydrogen to, can be purchased, which is the easiest way to go. There are suppliers in the resources list in the Appendix.

You can also do it yourself and buy your own hydrides, a bottle, and attendant fixtures. If you are interested in this approach there is more information available in *The Solar Hydrogen Chronicles,* edited by Walt Pyle.

When purchased, hydride bottles are initially activated by charging with hydrogen for extended periods of time, and/or at higher temperatures and/or at higher pressures. The particular process depends on the nature of the alloy used. It is common practice for the supplier to do the initial charge, however you can activate the bottle yourself.

The hydride bottle shown here is an aluminum bottle with an iron/titanium hydride. It has a 720 liter hydrogen capacity, is about 4.5 inches wide, about 10" high and weighs 12.6 pounds. This bottle has a female Swagelok QM series quick-connect fitting and a safety relief valve. The tank is designed to hold hydrogen up to pressures of about 500 psig.

The manufacturer (HCI) of this particular hydride bottle has several different hydride formulas available which they designate as A, L, M, and H. The bottle in the photos contains this manufacturers alloy A which is considered a nonflammable solid

Hydride bottle

267

when discharged. This means that it can be shipped without hazmat charges. The manufacturer will initially charge the bottle, then de-charge and ship it to you. The L, M, and H alloys that they offer are considered flammable solids and must be shipped as hazardous materials with the extra hazmat charge.

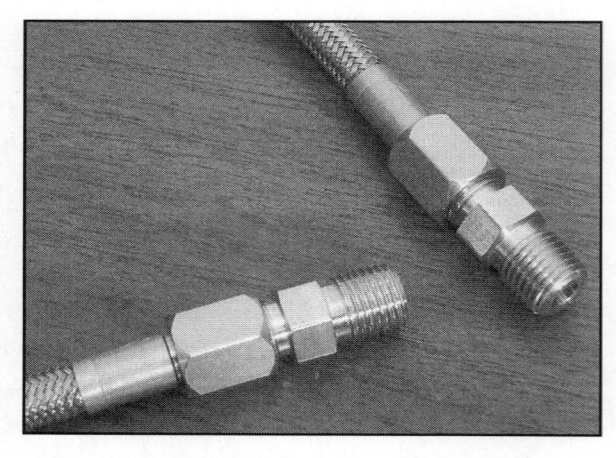

Braided hose with male swivel connects

This particular iron/titanium alloy requires a pressure of 400 psig to charge effectively. Lower pressures can be used to charge the alloy, but it will not charge to its full storage capacity.

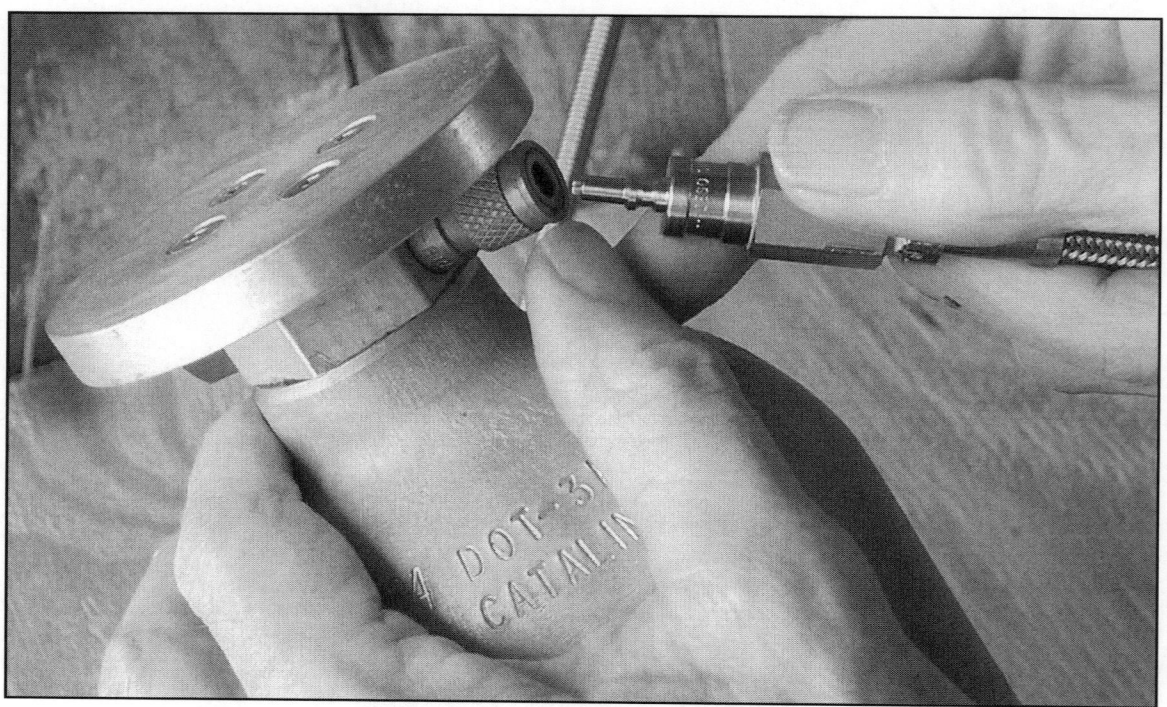

DESO stem used to connect to hydride bottle

Hydrides for Hydrogen Storage

Charging the hydride bottle

Charging is a simple procedure. You need a gas source that can deliver 400 psig and a regulator that can deliver 400 psig at the outlet. Most hydride bottles are filled from commercial high pressure hydrogen cylinders.

The cylinder regulator is connected to the hydride bottle by a flexible stainless steel hose with a ⅛" male NPT on one end and a ⅛" NPT male fitting on the other.

1. On one end of the hose, put a Swagelok quick-connect stem with a female NPT end that fits onto the hose. I prefer to use the DESO (double shut off valve) type stem rather than the SESO (single shut off valve) as it retains the hydrogen in the line and does not admit air contaminant back into the line connected to the regulator. This DESO stem fits on to the Swagelok quick-connect DESO body on the hydride bottle. Depending on the regulator you choose, the ⅛" NPT male end of the hose will either fit into the port on the regulator, or the regulator will have a ¼" NPT female outlet. If it has a ¼" NPT outlet, use a ⅛" female x ¼" male adapter to connect to the regulator.

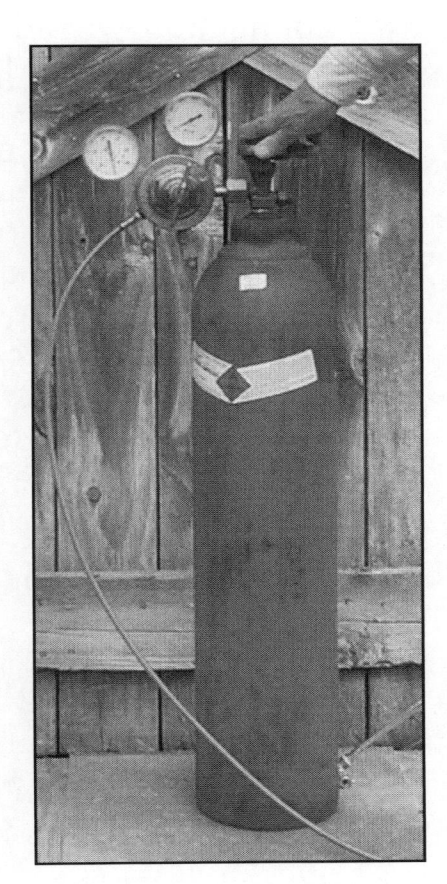

Open hydrogen cylinder valve

Practical Hydrogen Systems

2. Purge the hose and regulator of air. Open the cylinder wheel slowly up to full cylinder pressure and then open regulator valve to around 20 psi. Grab the quick-connect DESO stem that is attached to the hose and push down on the center of the stem inside the connection with a stick or small steel rod, to allow the gas to escape. Do this for a few seconds at a time for about ten times. This should be done outdoors so the gas can dissipate and not collect in a confined space. The best way to do this is to add another hose on to the regulator with a on/off valve and use this as a purging valve.

Purging hose and regulator though DESO stem

3. Connect regulator to hydrogen cylinder if not already in place.

4. To charge the hydride bottle, connect the hydride bottle to the cylinder regulator via the DESO stem connection.

5. The cylinder wheel is already opened to full cylinder pressure, so the next step is to open the regulator valve to delivery pressure of 400 psi.

6. When the cylinder charges it will become very warm. When it is done charging, it will cool down. You can speed up the charging process a bit by placing the hydride bottle in a container of cold water.

Hydrides for Hydrogen Storage

7. When the charging process is finished, turn off the hydrogen cylinder wheel valve and the regulator valve. You can leave the flexible hose attached to the cylinder regulator for the next fill. At each filling, purge the hose and regulator either as described above, or by releasing the vent hose on the regulator if you have added one.

8. Remove the quick-connect stem on the hose from the hydride bottle.

Fill the hydride bottle

Using the hydride bottle

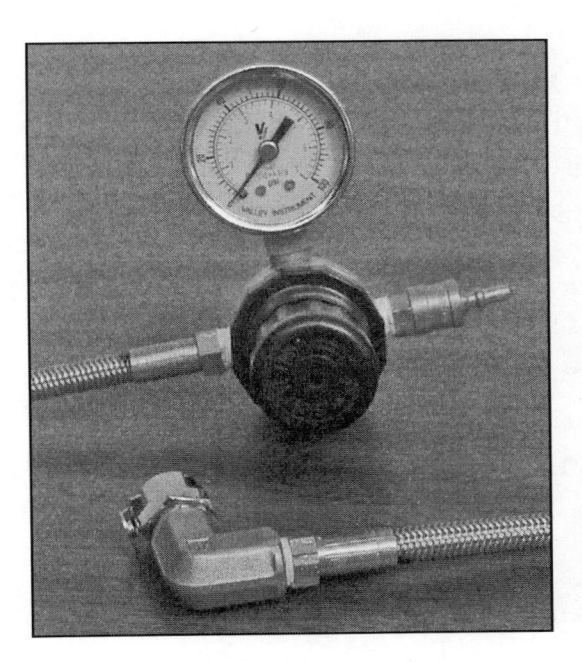

Hydride bottle regulator

To use the hydride bottle, connect a braided hose with DESO stem and a small regulator to the bottle. The other end of the hose will have the connection that goes to your particular hydrogen device, such as fuel cell stack.

Most hydrogen fueled devices use around 2 to 6 psi, and not over 15 psi, although there are some that use 25 psi. Whatever your application, you must use a regulator that

271

Connect regulator to hydride bottle

has the precision needed to work in the correct pressure range for your application.

Hydrides are interesting to work with. They can save space when a small footprint is desired. Applications such as robotics, mobile and remote locations can benefit in some circumstances from the use of hydride storage.

The heat generated as the alloy adsorbs the hydrogen, and the cooling generated by desorption as the hydrides release hydrogen, has potential for some interesting cogeneration applications, which would increase system efficiency where such applications are employed.

Hydrides for Hydrogen Storage

Gas cylinder and hydride storage resources

Description	Supplier	Part #	Quantity
Swagelok quickconnect stem with ⅛" NPT female connection/SS, DESO	Cambridge Valve & Fitting, Inc.	SS-QM2-D-2PF	1
PTFE braided hose w/type 304 SS braid rigid male x male fitting, 72" long, ³⁄₁₆" ID, 3000 psi	McMaster-Carr	4468K471	1
Hydride bottle, 720 liter	Hydrogen Components Inc.	CL-720	1
Regulator for hydride bottle	Fuel Cell Store	595618	1
Cylinder regulator/ 400 psi delivery	Fuel Cell Store	561213	1
Hydrogen cylinder	Local gas cylinder distributor such as Merriam Graves		
Connectors, general	Parker; Cambridge Valve and Fitting; Colder Products	See catalogs	
Metal alloys for hydride bottles	Alfa Aesar, Shieldalloy Metallurgical Corporation, GFE	See catalogs	
Hydride bottles	H Bank Technology Inc.	See catalog	
Pressure bottles for storage of hydrides	Aerocon Aerospace	See catalog	
Filter material for pressure bottles for hydride storage	Alfa Aesar	See catalog	

Parts numbers and suppliers have been provided for your convenience; however, suppliers may go out of business and parts numbers may change. All parts listed are available from multiple suppliers.

Commercial Fuel Cell Units

Mass flow controllers

Some fuel cell units and other applications may require very accurate measurement and/or control of gas flow.

This can be done with an MFC (mass flow controller). An MFC will measure the flow of gas, as well as control the flow when set up with attendant circuitry. The particular model shown in the photo has a 0-5 input/output and is powered by 15 volt DC source. It has a maximum operating pressure of 1500 psi and the valve is NC (normally closed).

We use MFCs for testing fuel cells and fuel cell stacks that we build in the lab, as well as for other experiments where a measured accurate flow is needed.

Mass flow controller

When ordering MFCs, you have to indicate the range of flow rate desired as well as the gas so that the MFC can be calibrated at the factory.

Purchasing fuel cell units

If you are not going to build your own fuel cell units, you can purchase a new or used one. New units are costly but have the advantage of warranty and customer service.

Used units need to be carefully considered before purchasing. Research the unit and find out what the runtime life is for that particular model. Then, enquire about how many runtime hours the unit has on it.

Be sure to get the manual with the unit. If the unit doesn't come with the manual, check to see if you can get one from the manufacturer or other secondary source. I do not recommend buying a unit without having access to the operator's manual.

Some fuel cell units are quite simple and consist of a simple stack with few attendant fittings. Others have complicated electronics and can be configured in a variety of different ways. It is a burden and can become a safety issue if your knowledge does not match the sophistication of the device. Newer models of the same series often have different operating parameters, so do not depend on an operating manual for a newer model of the same series being applicable for an older model.

As an example, I have an older unit that has been superseded by a newer model. The newer model has very nearly the same readouts and functions as the older model. However, there are a few differences, for instance, the gas supply pressure requirements. If I were to use the newer manual to operate the older model, I would apply far too much gas pressure to the unit. This could blow out a few MEAs, which would destroy part of the unit, leading to costly repairs, and downtime.

Find out what, if any, technical support might be available from the manufacturer, and what the availability of replacement parts is.

For some units, parts may be no longer available. This may or may not be of importance for you, but should be a consideration when pricing used units. You should find out what the manufacturer's stated run time life is for the unit you are considering; and keep in mind that some fuel cell units are designed to be backup power supplies and are not intended for constant running, full time power delivery.

Consider the cost of shipping the unit. You can get a very good deal on some used units, but the shipping may be very high if it is a heavy unit,

Commercial Fuel Cell Units

and if it is being shipped a considerable distance. Check out the shipping cost in total and be sure to include shipping insurance in your costs. Also consider what additional equipment will be needed to operate the unit.

In general, try to discern what condition the equipment is in, how much runtime life it has left, whether or not you can get replacement parts, and from there calculate whether the unit is cost effective for your purposes. Cover all the bases that are important for you before committing to purchase. If you forget to ask critical questions before you buy, you could be stuck with a unit that might not work, or that is too costly to repair.

Fuel cell unit options

Fuel cell units come in a variety of configurations. The two basic options are plain fuel cell stacks and fuel cell systems.

Fuel cell stacks are manufactured to either be stand alone or to be installed in a particular fuel cell system. Stand-alone stacks are generally used for research, teaching or OEM development. Fuel cell systems are basically manufactured to be backup power systems.

Stacks are simple affairs with no attendant electronics for the most part. They consist of a layered arrangement of fuel cells, gas delivery and exhaust ports and perhaps a fan that provides air flow for the unit.

Most fuel cell systems have a CPU/controller board with options for remote hardline or

Fuel cell unit in operation

277

wireless control by logic or computer interfaces. Other types are a little less exotic, but usually have some type of micro controller. Safety functions such as leak alarms and shutdown are included in most systems.

Such bells and whistles are great when everything works the way it should. If not, then you have to understand the system in order to get it back in working order. This means having the knowledge to test and analyze; and of course you also need to be able to get replacement parts, or fabricate your own. If you have a basic understanding of micro controllers, sensors and the basics of DC power units you will have a head start.

Most of a non-working system's parts will probably be in good working order. Often there is only one item that is throwing everything off. The good news is that most manufacturers use standard, readily-available parts. You can also strip out the automatic controls if they don't work properly and use the fuel cells with manual controls or design your own simple control unit.

On the other hand, if the price is right, a unit that is not in working order can be used to study the anatomy of fuel cell systems; and then perhaps be repaired. Or, you could scavenge the parts to use in a system of your design, or redesign the old system to work with newly fabricated parts.

The only really serious problem with an unknown system is if the fuel stack is pretty well spent. Contact the manufacturer before purchasing a unit to find out if replacement MEAs and other stack components are available, and what the costs are.

Basic system inspection

When you receive your unit be sure to inspect the surfaces for damage. Surface damage can affect operation. For instance, one unit we received had a dent in the door that prevented it from closing completely. The door was connected to a switch so that the unit would not turn on unless the door was closed. The door had to be hammered back into shape to correct the situation.

Commercial Fuel Cell Units

Read the manual before you first open up the fuel cell unit and start poking around. There may be special instructions for accessing certain parts of the unit. Also note if the unit is dirty or not. If it is, vacuum and wipe down the unit as well as possible. Many units have air ducts, filters and fan blades that should be inspected and cleaned on a regular basis.

Fuel cell cartridges visible below control panel

Check breaker switches

Some units have batteries to run components like fans in the unit. They are kept charged by the fuel cells. When the unit is shipped, the wire connectors are removed from the battery supply, and the breakers in the system are turned to the off position if the batteries are left in the unit when shipped.

If you're buying a used unit with batteries in it, ask what condition they are in. If they are old and not taking a charge, you can reduce shipping costs by having the seller remove the batteries, which are quite heavy.

279

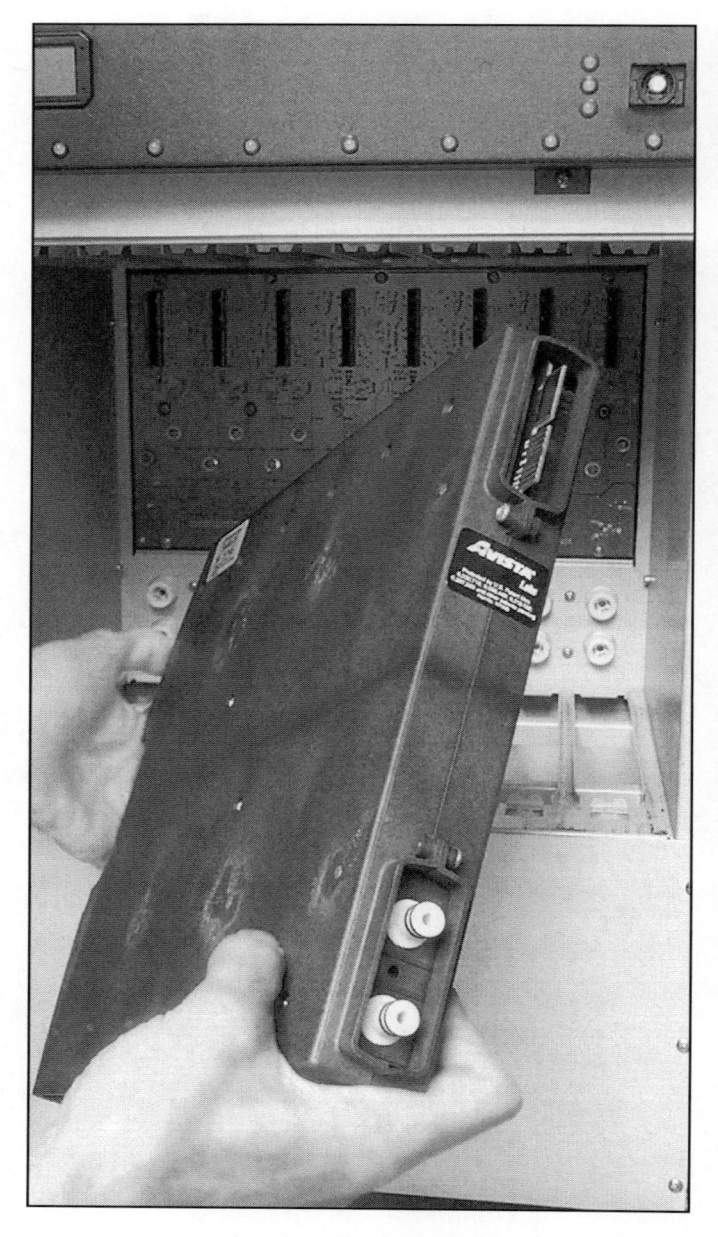

Fuel cell cartridge

In this case, you will have to order the appropriate batteries to install when you receive the fuel cell unit.

If the fuel cell unit comes with rechargeable batteries, put them on a charger and see how they charge up. Many units will spend hydrogen at startup to recharge the internal batteries before they will output power. This can take 20 minutes or more depending on the condition of the battery. Most systems require that batteries be at a certain level of charge. It's best to start out with fresh batteries for a unit, and not depend on used batteries that come with the equipment.

Some units have optional internal batteries, and can operate without them, and are solely used for charging an external battery bank. These different configurations will be noted in the operator's manual. Most units use sealed lead acid batteries that require chargers specifically designed for sealed batteries.

Of course, you can always use the fuel cell unit to charge up the batteries. The point is that if the batteries are old they should be replaced, so that you do not waste gas every time the unit is turned on, charging up batteries that are not taking the charge well.

280

Commercial Fuel Cell Units

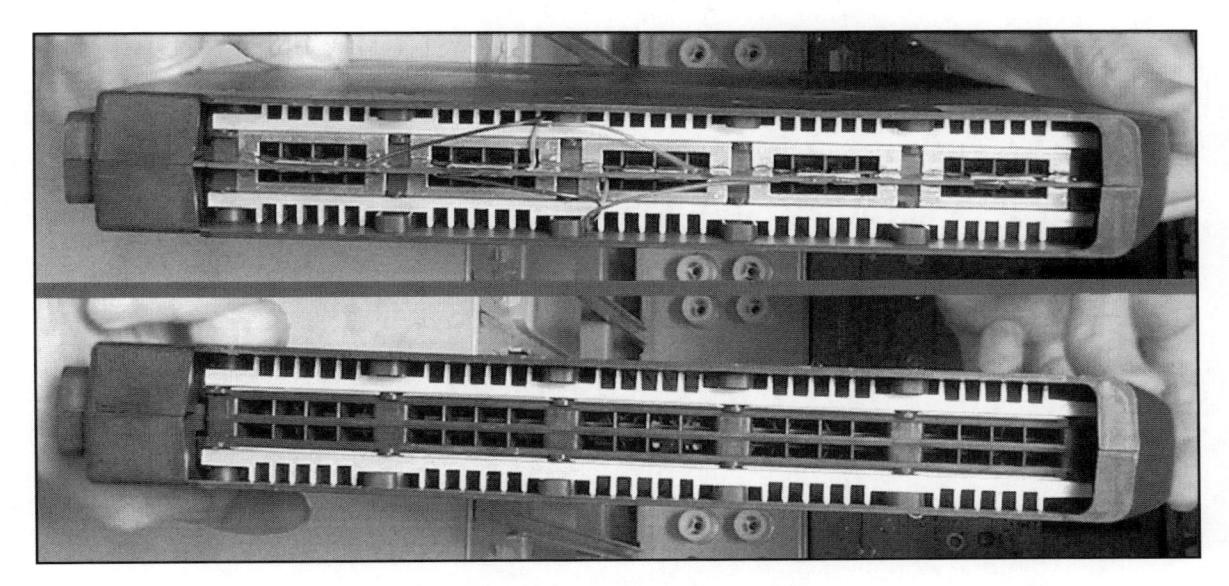

Fuel cell cartridge details

Some units have a temperature sensor attached by soft glue to one of the internal batteries for temperature compensation during charging. When you change batteries, carefully peel the sensor off the old battery and apply to the new battery. If the sensor needs to be replaced, most electronic supply houses carry them.

Locate the fuses and check to see if any need to be replaced. Check the general electrical components, connections and electronics in the system. If there are loose wires, find out why they are not connected and reconnect them if they should be. Loose wiring may be there to connect peripherals that were not included in the unit you have. We received a unit that had mysterious dangling wires. After a lot of head scratching, we discovered that they were for a heating pad peripheral that was not included with the unit. The heating pad is used to warm the fuel cells in cold operating temperatures.

Most fuel cell units have an error code display alert that will tell you what is wrong with the system if it is not working properly, and an automatic shutdown sequence if there are gas leaks or other problems that require immediate action.

Practical Hydrogen Systems

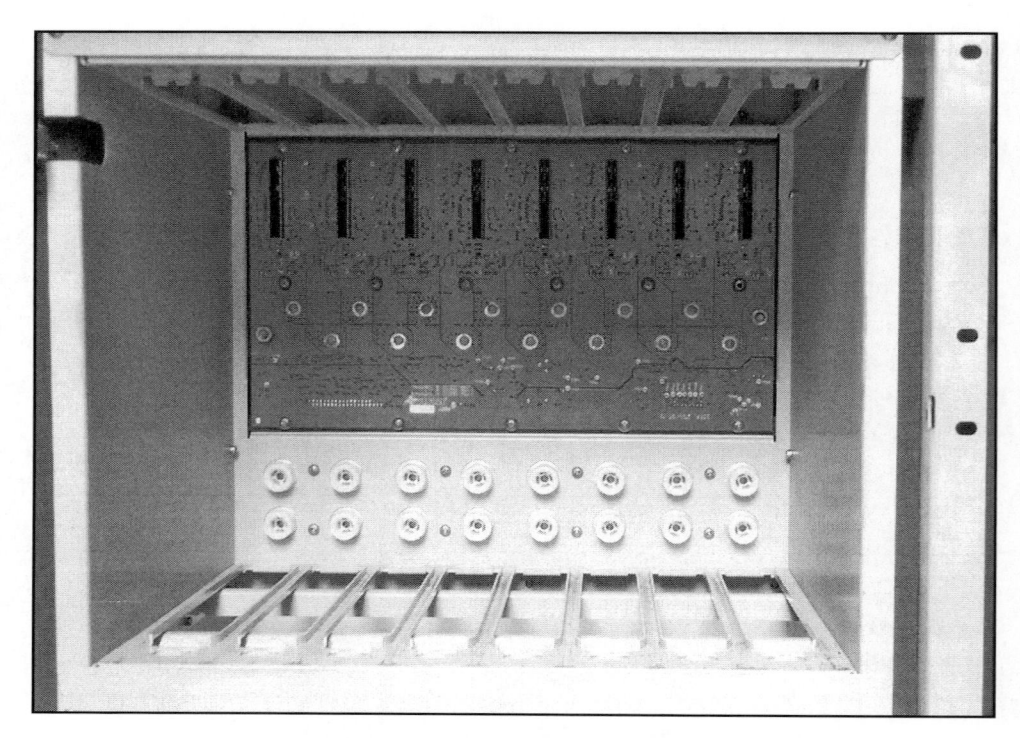

Inside unit with fuel cell cartridges removed

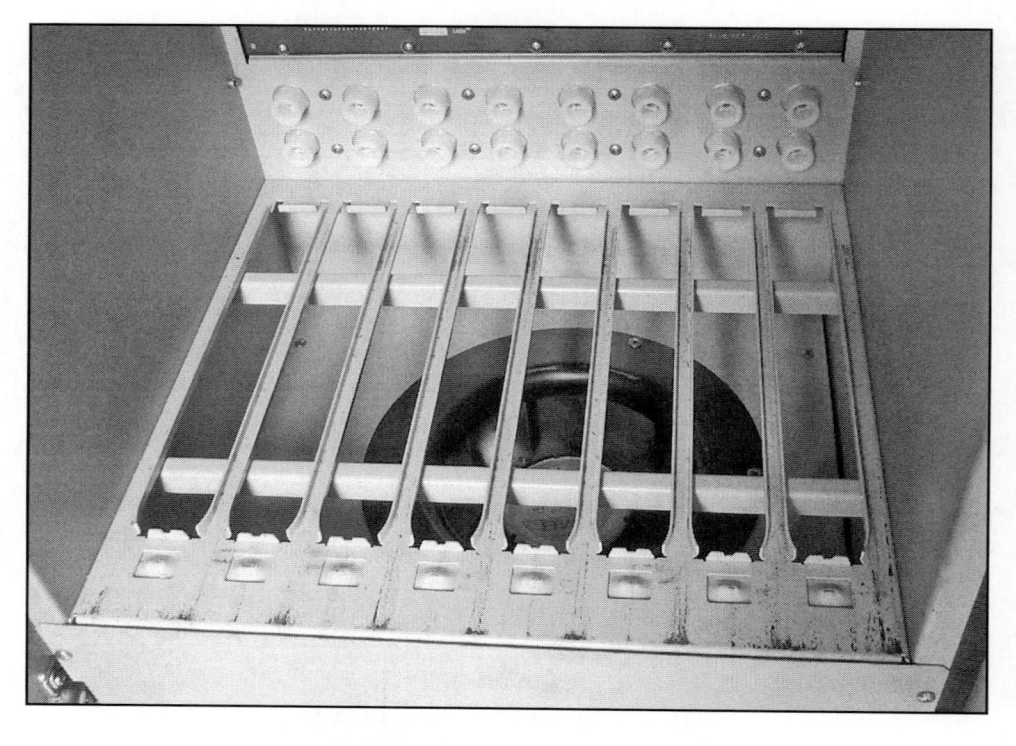

Check the fan and gas ports

Commercial Fuel Cell Units

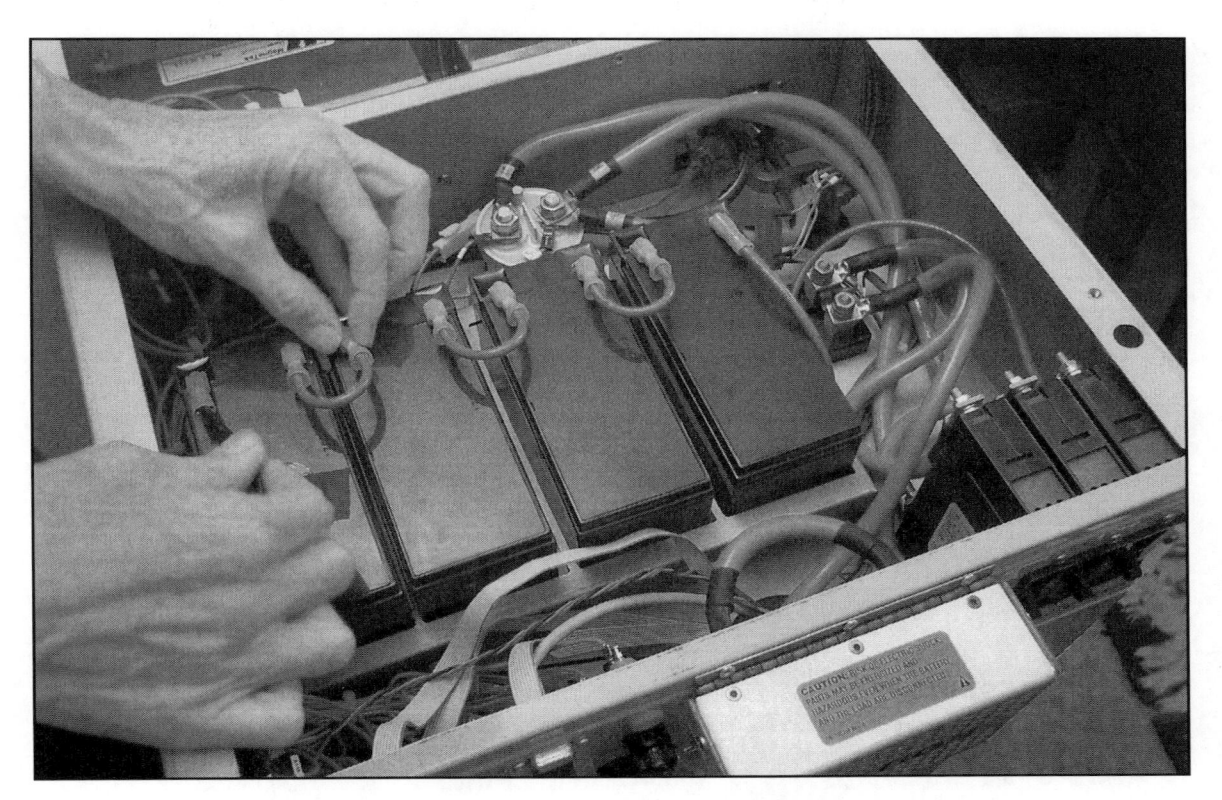

Check battery wires and other electrical connections

Unit control board

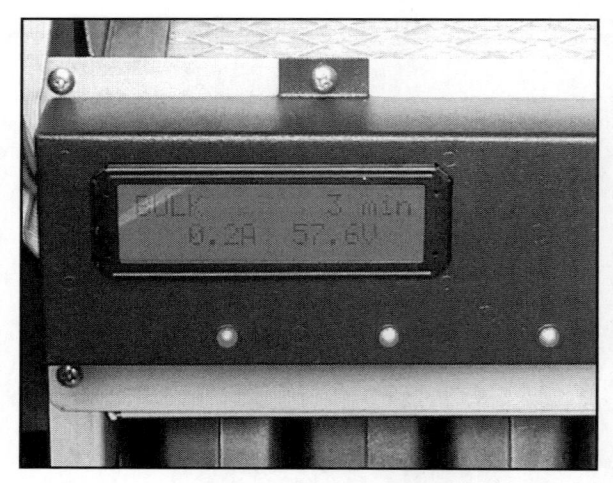

Digital readout of operating status

Check the fuses

Connecting to gas inlet. Discharge tube shown on right.

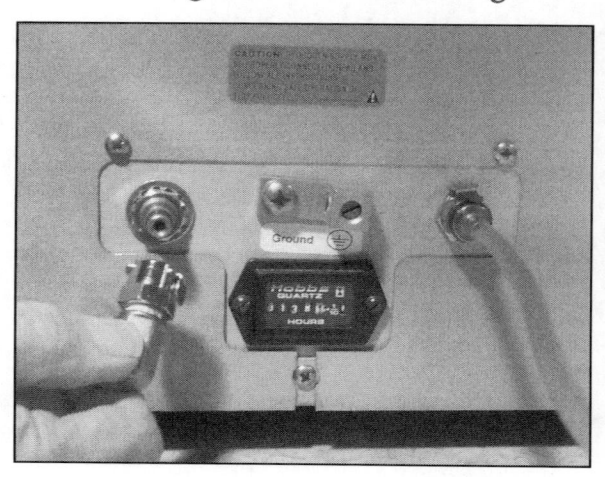

The fuel cells in these systems can vary widely as to their particular configuration. The unit shown in the photos has a sophisticated system of replaceable cartridges that each contain a number of fuel cells. The cartridges can even be replaced, if necessary, while the unit is running.

If you purchase simple stacks rather than a fuel cell system you will have a less complicated configuration to work with. Stacks may or may not have attendant fans and temperature control devices. They are quite simple affairs and, in general, you add the attendant electronics and peripherals.

With both fuel cell systems and stacks, the replacement costs for MEAs (membrane electrode assemblies) is a concern. If you can not get replacements for MEAs, then the fuel cells are of limited value.

There are many types of fuel cell systems and stacks available on the market available from current manufacturers. You can find information about these units on the internet.

Appendix

Suggested Reading

If you do not have an electronics or electrical background, the following titles will help you understand the principles behind the circuitry used in this book.

Getting Started In Electronics by Forrest Mims III, Master Publications.

Electronic Sensor Circuits And Projects by Forrest Mims III, Master Publications.

Beginners Guide To Reading Schematics by Robert J. Traister and Anna L. Lisk, Tab Books.

Practical Electronics For Inventors by Paul Scherz, McGraw-Hill.

A handy reference for pipe, tube, fitting and flange information is:

IPT's Pipe Trades Handbook by Robert A. Lee, IPT Publishing.

One serial publication of interest is *Sensors* magazine.

Suppliers' catalogs are important educational tools and I recommend these as a start:

Dwyer Instruments, Inc.

Cooper Crouse-Hinds

Automation Direct

Thermadyne/Victor

Advanced Specialty Gas Equipment

Specialty Gas Equipment

United States Plastic Corp.

Small Parts Inc.

Micro-Mark

Little Machine Shop

McMaster-Carr

Mead Fluid Dynamics

Allied Electronics

Cole-Parmer

Parker Hannifin Corp.

Most of these suppliers have a web presence and a catalog online, however I recommend ordering print catalogs, which contain some very good technical information and are excellent reference materials

There are many more suppliers and you can use the web to research and easily find the component or information that you will need.

Basic high school chemistry and physics texts can be very helpful for understanding electrochemistry, electricity, and gas laws in relation to hydrogen generation and storage.

Suppliers and Other Resources

This list of suppliers has been provided for your convenience; however, suppliers may go out of business. All parts listed in this book are available from multiple suppliers.

Supplier	Notes
McMaster-Carr	Too many items to list
Radio Shack	Electronic and electrical supplies
American Science & Surplus	Pressure regulators and other items of interest
All Electronics	Electronic and electrical supplies
Surplus Sales of Nebraska	Electronic and electrical supplies
Ocean State Electronics	Electronic and electrical supplies
Dwyer Instruments	Flow meters
KNF Neuberger	Explosion proof pumps
Hydrogen Components Inc.	A variety of hydrogen related components, connectors, hydride storage devices, etc.
Aero Tech Laboratories	Hydrogen storage, including pillow bags

Practical Hydrogen Systems

Supplier	Notes
Scott Specialty Gases	Pressure regulators, material compatibility tables, etc.
Aerocon Aerospace	High pressure cylinders, valves and more
Neodym	Gas detectors
Western Enterprises	Regulators, connectors, flashback arresters, etc.
Herbach and Rademan	Regulators, gauges, etc.
Micro-Mark	Hobby milling machines, routers, end mills, and lots of other useful stuff
Small Parts Inc.	A variety of small parts for fuel cell builders
Fuel Cell Store	Hydrogen equipment
H-Ion Solar	Gas processing systems and components for solar hydrogen systems
CGA - Compressed Gas Association	Information about fittings and other apparatus for gas systems
Alfa Aesar	Catalytic platinum/alumina pellets and other items. MSDS for hydrogen, oxygen, KOH.
Aquatic EcoSystems, Inc.	Tanks, tubing and other items of interest

Appendix

Supplier	Notes
Cole Parmer	Tubing, valves, connectors, pumps and much more
AHA - American Hydrogen Association	A good organization for the do it yourself enthusiast to join
Currents: Renewable Energy Information	Renewable energy hydrogen/solar
Surplus Center	Toggle switches, and other electrical components
The Electrochemical Society Inc.	Information
Automation Direct	Sensors
Allied Electronics	Electronic supplies
ITT Industries/Flojet	Pumps
Lindco Inc.	Valves, cylinders, etc.
Jameco	Electronics
Marlin P. Jones	Electronics
Convergence Tech, Inc.	Belfry Bat Detector

Practical Hydrogen Systems

Supplier	Notes
Swagelok/Cambridge Valve & Fitting Inc.	Gas processing components-fittings, valves, hoses, tubing, etc.
Generant	Valves
Lincoln Composites	Lightweight high pressure storage tanks
Sensors Magazine	Sensor information
H Bank Technology Inc.	Hydride storage
Brooks Instrument Div, Emerson Electric Co.	Flow controllers
Colder Products Co.	Quick couplings and fittings
RTI Alloys	Iron/titanium alloys
Servometer	Bellows
Mini-Flex Corp.	Bellows
Engineering Tips	Information resources
Motion Net	Information resources
Mat Web	Materials property data
Knovel	Scientific and technical data

Appendix

Supplier	Notes
Martindale's Reference Desk	General reference and on line calculators-such as gas law calculators and tutorials, electronics/electrical and everything else
Lessons In Electric Circuits by Tony Kuphaldt	Free e-books on electricity and electronic fundamentals – excellent
LM 317 Calculator by Mike Putnam	Calculator for DC voltage supplies
Voltage divider calculator by Bill Bowdon	Voltage divider calculator
United States Plastic Corp.	Plastic tanks, tubes, and more
Advanced Specialty Gas Equipment	Gas equipment and accessories
Specialty Gas Equipment	Gas equipment and accessories
Thermadyne/Victor	Regulators, gas apparatus
Circuit Specialists Inc.	Electronics
Mead Fluid Dynamics	Pneumatic components
Chemistry Store	Potassium Hydroxide (KOH)

Supplier	Notes
Fuel Cells Org.	Fuel cell systems and stack data
DRMS - US Defense Reutilization and Marketing Service	Surplus fuel cell units
Clean Fuel Cell Energy	MEAs and electrode components for fuel cells

Standards Information Resources

Organization	Notes
Alfa Aesar	MSDSs for hydrogen, oxygen, KOH
Cooper Crouse-Hinds	Hazardous area equipment information
ASTM - American Society for Testing Materials	Standards
ASME - American Society of Mechanical Engineers	Standards
NFPA - National Fire Protection Association	Standards
ANSI - American National Standards Institute	Standards
OSHA - US Occupational Safety and Health Association	Standards
ISO - International Organization for Standardization	Standards
SAE - Society Of Automotive Engineers	Standards
ECS - The Electrochemical Society	Standards

Organization	Notes
DOE - US Department of Energy	Standards
NREL - US National Renewable Energy Laboratory	Standards
IEC - International Electrotechnical Commission	Standards
IEEE - Institute of Electrical and Electronics Engineers	Standards
ISA - Instrumentation Systems, and Automation Society	Standards
DOT - US Department of Transportation	Standards

Titles by Phillip Hurley from
Wheelock Mountain Publications:

Solar II

Build Your Own Solar Panel

Build a Solar Hydrogen Fuel Cell System

Practical Hydrogen Systems

Build Your Own Fuel Cells

The Battery Builder's Guide

Solar Supercapacitor Applications

www.practicalhydrogen.com

Wheelock Mountain Publications
is an imprint of

Good Idea Creative Services
324 Minister Hill Road
Wheelock VT 05851
USA

Build Your Own Fuel Cells

by Phillip Hurley

The technology of the future is here today - and now available to the non-engineer! *Build Your Own Fuel Cells* contains complete, easy to understand illustrated instructions for building several types of proton exchange membrane (PEM) fuel cells - and, templates for 6 PEM fuel cell types, including convection fuel cells and oxygen-hydrogen fuel cells, in both single slice and stacks.

Low tech/high quality

Two different low-tech fuel cell construction methods are covered: one requires a bandsaw and drill press, and the other only a few hand tools. Anyone with minimum skills and tools will be able to produce high quality fuel cells from readily obtainable materials - contact info for materials suppliers is included.

Electrolyzers and MEAs

Build Your Own Fuel Cells includes a detailed discussion of building a lab electrolyzer to generate hydrogen to run fuel cells - and templates for the electrolyzer. Also covered is setting up a PV solar panel to power the electrolyzer, and experimental low-tech methods for producing membrane electrode assemblies (MEAs - the heart of the fuel cell).

Build Your Own Fuel Cells, 221 pages, over 140 B&W photos and illustrations, including 39 templates.

Available in print from Amazon.com

www.buildafuelcell.com

Build A Solar Hydrogen Fuel Cell System
by Phillip Hurley

Learn how to construct and operate the components of a solar hydrogen fuel cell system: the fuel cell stack, the electrolyzer to generate hydrogen fuel, simple hydrogen storage, and solar panels designed specifically to run electrolyzers for hydrogen production. Complete, clear, illustrated instructions to build all the major components make it easy for the non-engineer to understand and work with this important new technology.

Featured are the author's innovative and practical designs for efficient solar powered hydrogen production including:

- ESPMs (Electrolyzer Specific Photovoltaic Modules) – 40 watt solar panels designed specifically to run electrolyzers efficiently;
- a 40-80 watt electrolyzer for intermittant power from renewable energy sources such as solar and wind;
- and, a 6-12 watt planar hydrogen fuel cell stack to generate electricity.

Any of these components can be ganged or racked, or scaled up in size for higher output. You'll also learn how to set up an entire gas processing system, and where to find parts and materials – everything you need for an experimental stationary unit that will give you a solid base for building and operating systems for larger power needs. There are even schematics for adapting conventional solar panels (BSPMs – Battery Specific Photovoltaic Modules) for efficient hydrogen production, and setting up hybrid (battery and fuel cell) PV systems.

Available in print from Amazon.com

www.solarh.com

Printed in Great Britain
by Amazon